SpringerBriefs in Public Health

More information about this series at http://www.springer.com/series/10138

John A. Quelch • Emily C. Boudreau

Building a Culture of Health

A New Imperative for Business

John A. Quelch
Harvard TH Chan School of Public Health
Harvard Business School
Boston, MA, USA

Emily C. Boudreau
Harvard Business School
Boston, MA, USA

ISSN 2192-3698 ISSN 2192-3701 (electronic)
SpringerBriefs in Public Health
ISBN 978-3-319-43722-4 ISBN 978-3-319-43723-1 (eBook)
DOI 10.1007/978-3-319-43723-1

Library of Congress Control Number: 2016948733

Printed on acid-free paper

This Springer imprint is published by Springer Nature
The registered company is Springer International Publishing AG Switzerland

Dedicated to the Robert Wood Johnson Foundation
"Our goal is to help raise the health of everyone in the United States to the level that a great nation deserves, by placing well-being at the center of every aspect of life."

Preface

Every company, knowingly or unknowingly, impacts public health, and it does so in four ways: through the healthfulness and safety of the products and services it sells (Consumer Health); through the efforts it makes to insure the safety and well-being of not only direct employees but also workers in its supply chain (Employee Health); through investments it makes to improve health and safety in the communities where it does business (Community Health); and through the impact of its operations on the environment, through carbon emissions and water use, for example (Environmental Health). In these four ways, every company lays down a population health footprint. The net impact of the footprint can—and should—be measured. A company that incorporates a Culture of Health in its mission and daily decision-making will not only seek to make its net impact on public health as positive as possible, but will also create business opportunities for itself in doing so.

In April 2016, a conference was convened at Harvard Business School, in partnership with the Harvard T.H. Chan School of Public Health (represented by Professor Howard Koh) and the Robert Wood Johnson Foundation (represented by Executive Vice President Jim Marks). The title of the conference was the same as this book: Building a Culture of Health: A New Imperative for Business. The 300 attendees included around 60 % from the private sector, 20 % from the not-for-profit sector and from government, and 20 % from academia. Panels discussed Consumer Health, Employee Health, Community Health, and Environmental Health. Concluding sessions addressed how to connect the dots, measure a company's overall population health footprint and implement a culture of health in a company. The consensus at the conference was that this is a useful starting point—there is much more work that must be done to fundamentally reframe how business thinks and acts in the realm of public health.

This book, and the examples of company best practices that are included, draw from the conference proceedings. As such, we are deeply grateful for the ideas generated by the participants and for the partnership of the co-sponsors that made the conference possible. The Harvard Business School Division of Faculty Research and Development funded the by-invitation-only conference; we thank Dean Nitin Nohria for his support of our cross-disciplinary and cross-sector initiative.

In addition, we wish to thank colleagues who reviewed earlier drafts of portions of the manuscript: Professors Jose Alvarez and Walter Willett (Community Health); Professors Kate Baicker and Robert Huckman (Employee Health); Professors Howard K. Koh and V. Kasturi Rangan (Community Health); and Professors Rebecca Henderson, Eileen McNeely, and Jack Spengler (Environmental Health).

Finally, we acknowledge the help and efficiency of Elaine Shaffer who worked closely with our editor at Springer, Janet Kim, to move our manuscript through to publication.

Boston, MA, USA John A. Quelch
Boston, MA, USA Emily C. Boudreau
June 2016

Contents

Chapter 1
Building a Culture of Health

Every company, large and small, impacted health in four main ways. First, through the healthfulness and safety of the products and services it sold. Second, through its attention to employee health and well-being in its work practices and benefits. Third, through contributions to the broader communities in which it operated. And, fourth, through the environmental impacts of its operations.

Consider, for example, ride-sharing companies. By 2016, Uber, Lyft, and other ride-sharing companies were growing rapidly, conveniently connecting passengers who needed rides with nearby drivers—all through simple apps on their passengers' smartphones. These companies were not "healthcare" companies in a traditional sense.

However, they affected health in both positive and negative ways. In 2015, Uber produced a report in partnership with the non-profit organization, Mothers Against Drunk Driving ("MADD"). The report highlighted how drunk-driving crashes fell 6.5 % among drivers under 30 in California markets after Uber launched; what's more, 93 % of people surveyed would recommend Uber "as a safer way home to a friend who had been drinking."[1] Chariot, a ride-sharing company announced in 2016, was created in an effort to make rides safer for women by allowing only female drivers and female passengers.[2]

Reprinted with permission of Harvard Business School Publishing
Building a Culture of Health, HBS No. 9-516-073
This case was prepared by John A. Quelch and Emily C. Boudreau.

[1] MADD, "New Report from MADD, Uber Reveals Ridesharing Services Important Innovation to Reduce Drunk Driving," *MADD website,* January 27, 2015, http://www.madd.org/media-center/press-releases/2015/new-report-from-madd-uber.html?referrer=https://www.google.com/, accessed March, 2016.

[2] WMAR staff, "Women-only ride sharing app offers Uber alternative," *ABC 2 WMAR Baltimore,* http://www.abc2news.com/news/in-focus/women-only-ride-sharing-app-offers-uber-alternative, accessed April, 2016.

J.A. Quelch, E.C. Boudreau, *Building a Culture of Health*, SpringerBriefs in Public Health, DOI 10.1007/978-3-319-43723-1_1

While the ride-sharing apps provided new—and potentially safer—ways for consumers to travel, their health effects were not all necessarily positive. Around the same time, Uber found itself in the middle of an employee vs. independent contractor[3] debate, prompting questions about whether Uber drivers deserved employee benefits such as health insurance. Having health insurance was directly related to health-care access, and those with health insurance enjoyed better health outcomes.[4]

Because Uber's drivers used their own cars to provide rides and created their own schedules, Uber considered its drivers to be independent contractors, rather than employees. Uber maintained that the company was "merely an app that connect[ed] drivers and passengers—with no control over the hours its drivers work[ed]."[5] However, in 2015, a California court ruled that a prior Uber driver was in fact an employee.[6] The issue remained contentious, as the ruling did not apply beyond the driver who initiated the case.[7] Complicating the employee health debate further, there were also potential health benefits that independent contractors enjoyed, such as the flexibility to decide when, where, and how long to work and the ability to work for multiple companies—potentially factors that could reduce stress.[8,9] Therefore, it was difficult to assess the overall impact that ride-sharing companies had on employee or contractor health.

Assessing their net impact on community and environmental health was equally challenging. In 2014, both Lyft and Uber announced that they were releasing new passenger-pooling ride options that offered riders cheaper fares if they shared their vehicle with other passengers traveling along a similar route. One *TIME* article stated, "multiple people sharing a single ride to a common destination is a simple act that has the potential to reduce CO_2 emissions, ease traffic, lessen fossil fuel dependency, reduce stress on commuters, and even drive down rents in dense cities."[10]

---

[3] For employees, businesses were required to withhold income taxes, withhold and pay Social Security and Medicare taxes, and pay unemployment tax on paid wages. Businesses did not have to withhold or pay taxes on payments to independent contractors, and they did not have to offer them benefits like health insurance, paid time off, or overtime. In effect, hiring independent contractors was often much less costly for employers than hiring employees.

[4] Benjamin D. Sommers, MD, PhD; Sharon K. Long, PhD; and Katherine Baicker, PhD, "Changes in Mortality After Massachusetts Health Care Reform: A Quasi-experimental Study," *Annals of Internal Medicine*, 2014;160(9):585–593, doi:10.7326/M13-2275, accessed April 2016.

[5] Mike Isaac and Natasha Singer, "California Says Uber Drive is Employee, not a Contractor," *The New York Times,* June, 17, 2015, http://www.nytimes.com/2015/06/18/business/uber-contests-california-labor-ruling-that-says-drivers-should-be-employees.html, accessed October, 2015.

[6] Mike Isaac and Natasha Singer, "California Says Uber Drive is Employee, not a Contractor," *The New York Times,* June, 17, 2015, http://www.nytimes.com/2015/06/18/business/uber-contests-california-labor-ruling-that-says-drivers-should-be-employees.html, accessed October, 2015.

[7] Mike Isaac and Natasha Singer, "California Says Uber Drive is Employee, not a Contractor," *The New York Times,* June, 17, 2015, http://www.nytimes.com/2015/06/18/business/uber-contests-california-labor-ruling-that-says-drivers-should-be-employees.html, accessed October, 2015.

[8] Ibid.

[9] Jeanne Sahadi, "When an independent contractor is really an employee," *CNN Money,* July, 16, 2015, http://money.cnn.com/2015/07/16/pf/independent-contractors-employees/, accessed October, 2015.

[10] Katy Steinmetz, "How Uber and Lyft Are Trying to Solve America's Carpooling Problem," *Time,* June 23,2015, http://time.com/3923031/uber-lyft-carpooling/, accessed April, 2016.

Ride-sharing might thereby improve the health of the surrounding communities and reduce fossil fuel emissions—a boon for environmental health.[11] Further, there was potential for ride-sharing companies to reduce the number of consumers who found it necessary to own car, which would positively impact the environment and conserve resources.

However, the overall impact of these services on both community and environmental health remained elusive, as there was little information on what alternative transportation consumers were substituting (e.g., personal vehicles, public transportation, and taxi cabs).[12] Though sharing vehicles had the potential to decrease traffic congestion and emissions, passengers switching from walking to their destinations or taking new trips that would not have occurred otherwise would boost CO_2 emissions.

While the net impact of ride-sharing companies on public health was hard to measure, the industry illustrated that all companies—even those that did not seem at all health-related—had significant, multifaceted impacts on public health. It was also clear that companies could have both positive and negative effects on health and most corporations did not fully understand the potential influence they held. The objective here is twofold: first, to highlight how corporations wittingly and unwittingly enhance and detract from public health; and second, to showcase why it is critical—now more than ever—for corporations to prioritize building cultures of health as part of their missions and values.

Background

Health was an issue of global importance. Health status at an individual level profoundly affected quality of life, and at a population level, it affected a wide range of socioeconomic issues, in turn impacting many facets of everyday life.[13] Research by the Harvard T.H. Chan School of Public Health and the John F. Kennedy School of Government had shown that good population health improved the economy in five ways: financial protection; education; productivity; capital investments; and the demographic dividend. (See Exhibit 1.1 for detailed descriptions of each; see Exhibit 1.2 for a framework connecting health and poverty alleviation.)

Exhibit 1.1: How Good Health Reduces Poverty

1. **Financial protection**: Removing financial barriers to access enables the use of health services when needed, and helps at-risk households avert impoverishing expenditures and poverty.

[11] Ibid.

[12] Ibid.

[13] Health Poverty Action, "Key Facts: Poverty and Poor Health," *Health Poverty Action website*, https://www.healthpovertyaction.org/info-and-resources/the-cycle-of-poverty-and-poor-health/key-facts/, accessed May 2016.

2. **Education**: The prospect of longer, healthier lives induces people to invest more in their human capital, as they are better able to realize future long-term gains in employment and income.
3. **Productivity:** Productivity is enhanced through contribution of better health to increased worker capacity, lower rates of absenteeism, and less workforce turnover.
4. **Capital investments**: Heightened longevity in lifespan and higher incomes mean people save more for retirement—boosting the economy-wide capital available for increased investments.
5. **The demographic dividend**: With the right conditions in place, changes in population age structure with growing and educated work force creates the opportunity for economic growth.

Source: Rifat Atun, Claire Chaumont, Joseph R Fitchett, Annie Haakenstad, Donald Kaberuka, "Poverty Alleviation and the Economic Benefits of Investing in Health," *Forum for Finance Ministers 2016*, accessed at https://cdn2.sph.harvard.edu/wp-content/uploads/sites/61/2015/09/L-MLIH_Health-economic-growth-and--development_Atun-and-Kaberuka_4-11-16.pdf, accessed May, 2016.

Exhibit 1.2: Connections Between Health and Poverty

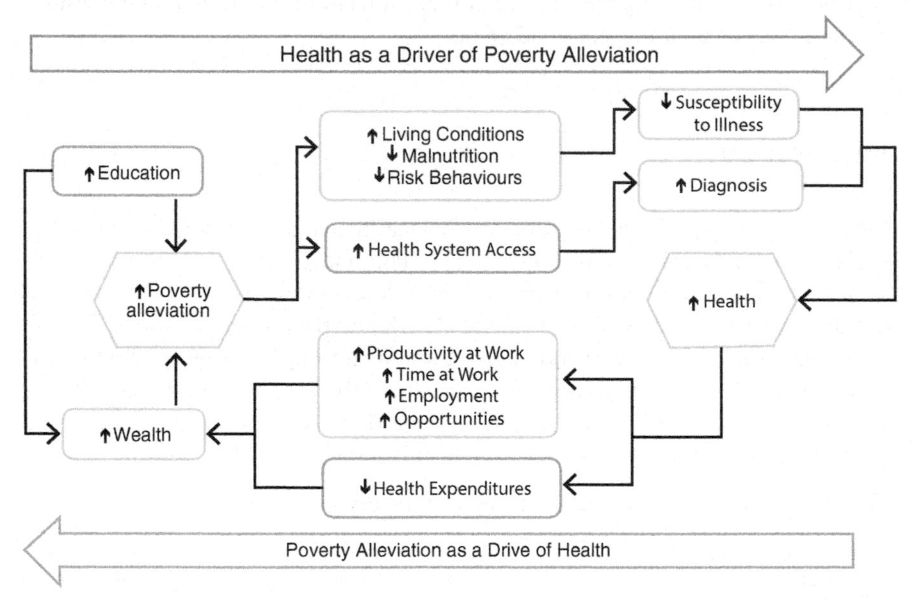

Source: Rifat Atun, Claire Chaumont, Joseph R Fitchett, Annie Haakenstad, Donald Kaberuka, "Poverty Alleviation and the Economic Benefits of Investing in Health," *Forum for Finance Ministers 2016*, accessed at https://cdn2.sph.harvard.edu/wp-content/uploads/sites/61/2015/09/L-MLIH_Health-economic-growth-and--development_Atun-and-Kaberuka_4-11-16.pdf, accessed May, 2016.

By the 2000s, international efforts to improve public health had achieved considerable success. Life expectancy at birth had increased worldwide—rising in the US, for example, from around 47 in 1900 to nearly 79 in 2013.[14,15] This dramatic increase was largely due to improvements in living conditions (e.g., sanitation, hygiene, housing, and education) and medical advances (e.g., vaccines and antibiotics).[16] Early to mid-life mortality was so reduced that, by the second half of the twentieth century, there was little room for further improvement.[17] A 2011 report by the National Institutes of Health stated, "The dramatic increase in average life expectancy during the twentieth century ranks as one of society's greatest achievements."[18]

Nevertheless, several significant public health challenges remained. These included:

Rising non-communicable diseases: Despite a decrease in the prevalence of infectious diseases, the prevalence of non-communicable diseases (e.g., cardiovascular disease, cancer, diabetes, and chronic respiratory disease) was increasing.[19] According to the Centers for Disease Control and Prevention (CDC), non-communicable diseases were responsible for more than 75 % of deaths worldwide in 2016.[20] Non-communicable diseases were largely caused by modifiable risk factors (e.g., tobacco use, alcohol consumption, unhealthy diet, and insufficient physical activity). The World Health Organization (WHO) estimated that 80 % of premature heart disease, stroke, and diabetes could be prevented.[21]

Aging populations: Populations around the world were rapidly aging as people continued to live longer.[22] By 2050, thc US population aged 65 and older was projected to reach 83.7 million, almost double the 43.1 million in 2012.[23] Growth in the elderly

---

[14] CDC, "Table 22: Life expectancy at birth, at 65 years of age, and at 75 years of age, by race and sex: United States, selected years 1900–2007, *CDC website,* http://www.cdc.gov/nchs/data/hus/2010/022.pdf, accessed April, 2016.

[15] Jiaquan Xu, Sherry L. Murphy, Kenneth D. Kochanek, M.A., Brigham A. Bastian, "Deaths, Final Data for 2013," *National Vital Statistics Reports,* volume 64, Number 2, February 16, 2016, http://www.cdc.gov/nchs/data/nvsr/nvsr64/nvsr64_02.pdf, accessed April, 2016.

[16] Rosabeth Moss Kanter, Howard Koh, Pamela Yatsko, "The State of U.S. Public Health: Challenges and Trends," HBS No. 316-001 (Boston: Harvard Business School Publishing, 2015).

[17] Ibid.

[18] National Institute on Aging, "Health and Aging: Living Longer," *U.S. Department of Health and Human Services,* 2011, https://www.nia.nih.gov/research/publication/global-health-and-aging/living-longer, accessed April, 2016.

[19] World Health Organization, "Noncommunicable diseases (NCD)," *WHO website,* http://www.who.int/gho/ncd/en/, access April, 2016.

[20] Centers for Disease Control and Prevention (CDC), "CDC Global Noncommunicable Diseases (NCDs)," CDC website, http://www.cdc.gov/globalhealth/healthprotection/ncd/, accessed May, 2016.

[21] World Health Organization, "Noncommunicable diseases (NCD)," *WHO website,* http://www.who.int/gho/ncd/en/, access April, 2016.

[22] World Health Organization, "Ageing and life-course," *WHO website,* http://www.who.int/ageing/en/, accessed April, 2016.

[23] Jennifer M. Ortman, Victoria A. Velkoff, and Howard Hogan, "An Aging Nation: The Older Population in the United States," *Census.gov website,* May 2014, https://www.census.gov/prod/2014pubs/p25-1140.pdf, accessed April, 2016.

population increased demand for health services and raised policy questions around the continuing financial viability of social security programs.[24]

Increasing costs: In 2013, health expenditures[25] totaled ~17 % of the US gross domestic product (GDP) and ~10 % of the worldwide GDP.[26] Healthcare costs increased throughout the 2000s due to faster-than-general inflation increases in the prices of drugs, medical devices, and hospital care.[27] As health costs increased, many in the US questioned the sustainability of government-funded health programs for the elderly and poor, Medicare and Medicaid, at current levels.[28]

Health and income disparities: Many, diverse factors affected the health of individuals and broader populations. Some of these elements were controllable by the individual, but many were not. Although average life expectancy had increased worldwide, it varied significantly across the world.[29] Further, these differences existed not only between countries, but also within them.[30] There was no biological reason for these differences and experts largely attributed them to social and environmental factors, such as income inequality and healthcare access disparities.[31]

According to the University of Wisconsin Population Health Institute, healthy behaviors determined 39 % of a population's health, while 12 % was attributable to health care and fully 50 % depended on social and economic factors.[32] Research by Robert Putnam showed an increasing opportunity divide between richer and poorer

---

[24] Ibid.

[25] Total health expenditure was the sum of public and private health expenditure. It covered the provision of health services (preventive and curative), family planning activities, nutrition activities, and emergency aid designated for health but does not include provision of water and sanitation.

[26] The World Bank, "Data: Health expenditure, total (% of GDP)." *The World Bank website,* http://data.worldbank.org/indicator/SH.XPD.TOTL.ZS, accessed April, 2016.

[27] Mike Patton, "U.S. Health Care Costs Rise Faster Than Inflation," *Forbes,* June 29, 2015, http://www.forbes.com/sites/mikepatton/2015/06/29/u-s-health-care-costs-rise-faster-than-inflation/#3665589f6ad2, accessed April, 2016.

[28] Kimberly Leonard, "Are Medicare and Medicaid Sustainable?" *U.S. News,* April 15, 2015, http://www.usnews.com/news/articles/2015/04/15/are-medicare-and-medicaid-sustainable, accessed April, 2016.

[29] Central Intelligence Agency, "The World Factbook," *CIA website,* 2015 Est., https://www.cia.gov/library/publications/the-world-factbook/rankorder/2102rank.html, accessed April, 2016.

[30] World Health Organization: Commission on Social Determinants of Health, "Closing the gap in a generation," *WHO website,* http://apps.who.int/iris/bitstream/10665/43943/1/9789241563703_eng.pdf, accessed April, 2016.

[31] The World Health Organization, Commission on Social Determinants of Health, "closing the gap in a generation: Health equity through action on the social determinants of health," 2008, http://apps.who.int/iris/bitstream/10665/43943/1/9789241563703_eng.pdf, p. 26, accessed November, 2015.

[32] Bridget C. Booske, Jessica K. Athens, David A. Kindig, Hyojun Park, Patrick L. Remingtom, "Country Health Rankings Working Paper: DIFFERENT PERSPECTIVES FOR ASSIGNING WEIGHTS TO DETERMINANTS OF HEALTH, *University of Wisconsin Population Health Institute,* February 2010, https://uwphi.pophealth.wisc.edu/publications/other/different-perspectives-for-assigning-weights-to-determinants-of-health.pdf, accessed April, 2016.

children in the US.[33] Children from higher income households typically had greater access to social support, healthcare, and extracurricular activities, which led to significantly better outcomes over time versus those for children from lower income households.[34]

Public and Social Sector Responses

Efforts to address these concerns were sponsored by governments, non-profit organizations, and international agencies. In 2000, the United Nations (UN) announced the Millennium Development Goals (MDGs), a set of eight goals that were focused on reducing extreme poverty worldwide between 2000 and 2015.[35] Five of the goals were directly related to health (see Table 1.1 for the MDGs). In 2015, Ban Ki-moon Secretary-General of the UN discussed the significant but incomplete outcomes from the MDGs, stating:

> The MDGs helped to lift more than one billion people out of extreme poverty, to make inroads against hunger, to enable more girls to attend school than ever before and to protect our planet. They generated new and innovative partnerships, galvanized public opinion and showed the immense value of setting ambitious goals…Yet for all the remarkable gains, I am keenly aware that inequalities persist and that progress has been uneven. The world's poor remain overwhelmingly concentrated in some parts of the world.[36]

Table 1.1 Millennium development goals

Goal	Description
1	Eradicate extreme poverty and hunger
2	Achieve universal primary education
3	Promote gender equality and empower women
4	Reduce child mortality
5	Improve maternal health
6	Combat HIV/AIDs, malaria and other diseases
7	Ensure environmental sustainability
8	Global partnership for development

Source: Adapted from UN, "Millennium Development Goals," *UN website,* http://www.un.org/millenniumgoals/bkgd.shtml, accessed April, 2016

[33] Robert Putnam, "Crumbling American Dreams," *The New York Times,* August 3, 2013, http://opinionator.blogs.nytimes.com/2013/08/03/crumbling-american-dreams/?_r=0, accessed May, 2016.

[34] Michael Jonas, "Opportunity gap," *CommonWealth Magazine,* October 13, 2015, http://commonwealthmagazine.org/economy/opportunity-gap/, accessed December 2015.

[35] UN, "The Millennium Development Goals Report 2015, *Un website,* http://www.un.org/millenniumgoals/2015_MDG_Report/pdf/MDG%202015%20rev%20(July%201).pdf, accessed May, 2016.

[36] Ibid.

Table 1.2 Sustainable development goals

Goal	Description
1	No poverty
2	Zero hunger
3	Good health and well-being
4	Quality education
5	Gender equality
6	Clean water and sanitation
7	Affordable and clean energy
8	Decent work and economic growth
9	Industry, innovation and infrastructure
10	Reduced inequalities
11	Sustainable cities and communities
12	Responsible consumption and production
13	Climate action
14	Life below water
15	Life on land
16	Peace, justice and strong institutions
17	Partnerships for the goals

Source: Adapted from UN, "Sustainable Development Goals: 17 Goals to Transform Our World," *UN website,* 2015, http://www.un.org/sustainabledevelopment/sustainable-development-goals/, accessed April, 2016

Building on the MDGs, the UN announced the Sustainable Development Goals (SDGs) in 2015, a "set of goals to end poverty, protect the planet, and ensure prosperity for all as part of a new sustainable development agenda."[37] Six of the goals were directly related to health, and all were indirectly related.[38] (See Table 1.2 for the SDGs.) In 2016, World Bank leaders met to discuss the importance of promoting healthier communities as a driver of economic growth.[39]

While the international interest in development was noteworthy, socioeconomic challenges (e.g., health, education, income, and employment) were often related to one another, meaning that any attempt to focus on public health could quickly become complicated. Additionally, there were concerns that the World Health Organization (WHO), a specialized agency of the UN, was underfunded and unable

[37] UN, "Sustainable Development Goals: 17 Goals to Transform Our World," *UN website,* http://www.un.org/sustainabledevelopment/sustainable-development-goals/, accessed April, 2016.

[38] United Nations, "Sustainable Development Goals," *Sustainable Development Knowledge Platform,* accessed at https://sustainabledevelopment.un.org/, accessed November 2015.

[39] The World Bank, "Overview," *The World Bank Website,* http://www.worldbank.org/en/topic/health/overview#1, accessed April, 2016.

to address the rising tide of new public health concerns, especially outbreaks of deadly viruses that could lead to pandemics.[40]

Around the same time, policy makers in the US undertook measures to improve health outcomes, access, cost, and equity. The 2010 Affordable Care Act (ACA) placed a new focus on disease prevention and wellness, and addressed the social determinants of health.[41] (See Exhibit 1.3 for the social determinants of health.) Despite high health care spending, the US continued to rank poorly versus other wealthy countries on standard measures of health system performance.[42] The ACA aimed to expand health insurance enrollment, improve quality of care, and simultaneously reduce health care costs.[43]

Exhibit 1.3: Social Determinants of Health

Economic Stability	Neighborhood and Physical Environment	Education	Food	Community and Social Context	Health Care System
Employment Income Expenses Debt Medical bills Support	Housing Transportation Safety Parks Playgrounds Walkability	Literacy Language Early childhood education Vocational training Higher education	Hunger Access to healthy options	Social integration Support systems Community engagement Discrimination	Health coverage Provider availability Provider linguistic and cultural competency Quality of care

Health Outcomes
Mortality, Morbidity, Life Expectancy, Health Care Expenditures, Health Status, Functional Limitations

Source: Harry J. Heiman and Samantha Artiga, "Beyond Health Care: The Role of Social Determinants in Promoting Health and Health Equity," *The Kaiser Family Foundation,* November 4, 2015, accessed at http://kff.org/disparities-policy/issue-brief/beyond-health-care-the-role-of-social-determinants-in-promoting-health-and-health-equity, accessed November, 2015.

In addition to policy reform, The Robert Wood Johnson Foundation (RWJF), the largest philanthropic organization focused on public health in the US, pushed for a

---

[40] Jason Gale and John Lauerman, "Ebola Spread Over Months as WHO Missed Chances to Respond," *Bloomberg,* October 17, 2014, http://www.bloomberg.com/news/articles/2014-10-16/who-response-to-ebola-outbreak-foundered-on-bureaucracy, accessed April, 2016.

[41] Rosabeth Moss Kanter, Howard Koh, Pamela Yatsko, "The State of U.S. Public Health: Challenges and Trends," HBS No. 316-001 (Boston: Harvard Business School Publishing, 2015).

[42] Ibid.

[43] Ibid.

new societal culture that embraced health and wellness. The organization termed its vision a "Culture of Health" and, with it, sought to address the underlying determinants of health, encourage cross-collaboration among different kinds of organizations, and improve health equity. (See Exhibit 1.4 for a description of RWJF's Culture of Health.) The term "Culture of Health" was used several years prior by both the American Hospital Association and Cigna, though both used it largely to refer to the culture within organizations that prioritized employee health programs.[44,45] RWJF expanded the term beyond employee health.

Exhibit 1.4: The Robert Wood Johnson Foundation "Culture of Health" (Excerpt)

There is no single definition, which means when America ultimately achieves a Culture of Health it will be as multifaceted as the population it serves. We believe an American Culture of Health is one in which:

1. Good health flourishes across geographic, demographic and social sectors.
2. Attaining the best health possible is valued by our entire society.
3. Individuals and families have the means and the opportunity to make choices that lead to the healthiest lives possible.
4. Business, government, individuals, and organizations work together to foster healthy communities and lifestyles.
5. Everyone has access to affordable, quality health care because it is essential to maintain, or reclaim, health.
6. No one is excluded.
7. Health care is efficient and equitable.
8. The economy is less burdened by excessive and unwarranted health care spending.
9. The health of the population guides public and private decision-making.
10. Americans understand that we are all in this together.

Source: Risa Lavizzo-Mourey, "Building a Culture of Health, *The Robert Wood Johnson Foundation,* 2014, http://www.rwjf.org/content/dam/files/rwjf-web-files/Annual_Message/2014_RWJF_AnnualMessage_final.pdf, accessed April, 2016.

In sum, there had been significant gains in public health, yet many challenges remained. Researchers had established clear relationships between health and a wide range of other societal issues (e.g., education, economic growth, and equality) and, as a result, governments, nonprofits, and international agencies placed a renewed focus on public health.

[44] American Hospital Association: 2010 Long-Range Policy Committee, "A Call to Action: Creating a Culture of Health," Chicago: American Hospital Association, January 2011, http://www.aha.org/research/cor/content/creating-a-culture-of-health.pdf, accessed May, 2016.

[45] Cigna, "Creating a Culture of Health," 2010, http://www.cigna.com/assets/docs/improving-health-and-productivity/837897_CultureOfHealthWP_v5.pdf, accessed May, 2016.

Business Involvement in Public Health

Despite these efforts, corporations remained largely uninvolved in public health initiatives even though there were potential business benefits to improving health (e.g., reduced costs, increased revenue, and improved reputation). Corporate involvement often came largely in the form of regulatory compliance. For example, in 2014, Mexico enacted a tax that amounted to a 10 % retail price increase for sugary soft drinks; manufacturers fought the new tax, but in the end, had to comply and responded by trying to make their product portfolios healthier.[46]

There were several indications that corporate attitudes towards health were beginning to change. In 2014, the Bipartisan Policy Center's (BPC) CEO Council on Health and Innovation produced a report outlining the ways in which businesses were combating public health challenges and how they could become more involved in the future.[47] The CEO Council's members included the CEOs of Verizon and Bank of America, among others. (See Table 1.3 for a full list of the CEO Council's members.)

The Health Enhancement Research Organization (HERO), a non-profit comprising members from private corporations, health systems, hospitals, wellness vendors, nongovernmental organizations, and foundations, stated that one of its 2015 research goals was "Employer-Community Collaboration in Improving Population Health: Studying community based projects where employers partner with communities in pursuit of the goal of improving the overall health of that population."[48]

Additionally, the Institute of Medicine (IOM) held roundtable discussions in 2014 and 2015 on the role of business in achieving population health priorities.[49] At the 2014 conference, participants discussed recent cases of corporate involvement in community health, but concluded that corporations could do more. The summary report noted that "many participants said they were encouraged by the examples of business engagement in population health that had been described throughout the

---

[46] Anahad O'Connor, "Mexican Soda Tax Followed by Drop in Sugary Drink Sales," *The New York Times,* January 6, 2016, http://well.blogs.nytimes.com/2016/01/06/mexican-soda-tax-followed-by-drop-in-sugary-drink-sales/, accessed April, 2016.

[47] CEO Council on Health and Innovation, "Building Better Health: Innovative Strategies from America's Business Leaders," *Bipartisan Policy Center,* accessed at http://www.healthinnovation-council.org/wp-content/uploads/2014/09/BPC_Health-Innovation-Initiative_Building-Better-Health-A-Report-from-the-CEO-Council-Sept-2014.pdf, accessed November, 2015.

[48] HERO, "The HERO Research Agenda–2015," accessed at http://hero-health.org/wp-content/uploads/2015/09/HERO-Research-Agenda-2015_final-2.pdf, accessed November, 2015.

[49] Institute of Medicine, "Meeting agenda: Applying a Health Lens II: The Role and Potential of the Private Sector to Improve Economic Well-Being and Community Outcome," June 4, 2015, *The National Academies of Science, Engineering, and Medicine,* accessed at http://iom.nationalacademies.org/activities/publichealth/populationhealthimprovementrt/2015-jun-04.aspx, accessed November, 2015.

Table 1.3 CEO Council members

Name	Company
Mark Bertolini	Aetna
Brian Moynihan	Bank of America Corporation
Scott Serota	Blue Cross and Blue Shield Association
Patrick Soon-Shiong	Institute for Advanced health and NantHealth
Alex Gorsky	Johnson & Johnson
Dominic Barton	McKinsey & Company
Muhtar Kent	The Coca-Cola Company
Lowell McAdam	Verizon Communications
Gregory Wasson	Walgreens

Source: The CEO Council on Health and Innovation, *Council website,* accessed at http://www.healthinnovationcouncil.org/, accessed November, 2015

workshop, but the sentiment was that there is much to be done to make these examples the rule rather than the exception."[50] The goals of the 2015 workshop included:

1. Explore what businesses can offer the movement to improve population health
2. Discuss areas of potential, as well as models for how businesses could impact the determinants of health
3. Provide a platform for discussing how to promote and support health in all business practices, policies, and investments[51]

The Sustainability and Health Initiative for NetPositive Enterprise (SHINE) at the Center for Health and the Global Environment within the Harvard T.H. Chan School of Public Health conducted research and created partnerships with companies to not only increase momentum for measuring and improving workforce well-being, but also create an evidence base that doing so had the potential to benefit society and the companies themselves.[52] For example, in 2013, SHINE worked with Johnson & Johnson to develop a workforce well-being index.[53] Metrics in the index went beyond typical measures (e.g., health care costs and wellness programs participation rates) and emphasized well-being (e.g., a psychological, social and physical state).[54] In addition, the index included measures for drivers of well-being at work, such as, opportunities for learning, autonomy, mastery, and social connection.[55]

---

[50] Theresa Wizemann, "BUSINESS ENGAGEMENT in BUILDING HEALTHY COMMUNITIES WORKSHOP SUMMARY," 2014, The National Academies Press, Washington, D.C., PDF accessed at http://www.nap.edu/catalog/19003/business-engagement-in-building-healthy-communities-workshop-summary, accessed November, 2015.

[51] Ibid.

[52] Karoline Barwinski, "Health and Well-being: A Business Imperative," *The Huffington Post,* April 4, 2016, http://www.huffingtonpost.com/karoline-barwinski/health-wellbeing-a-busine_b_9609230.html, accessed April, 2016.

[53] Ibid.

[54] Ibid.

[55] Ibid.

Eileen McNeely, founder and co-director of the program, wrote, "As human beings spend the majority of their lives in the workplace, we need better workplace structures to help them flourish and reverse the declining health in the U.S. and, increasingly, the world. We need to set a bold vision for companies to factor their impacts on health into all business decisions, and to act in ways that will protect us and our planet."[56]

In 2016, RWJF called for greater cross-sector collaboration and sought to engage the business community in its Culture of Health vision. In *From Vision to Action: A Framework to Mobilize a Culture of Health,* RWJF described an "Action Framework" that included four objectives: making health a shared value; fostering cross-sector collaboration to improve well-being; creating healthier, more equitable communities, strengthening integration of health services and systems.[57] (See Exhibit 1.5 for RWJF's Action Framework.)

Exhibit 1.5: Robert Wood Johnson Foundation Action Framework

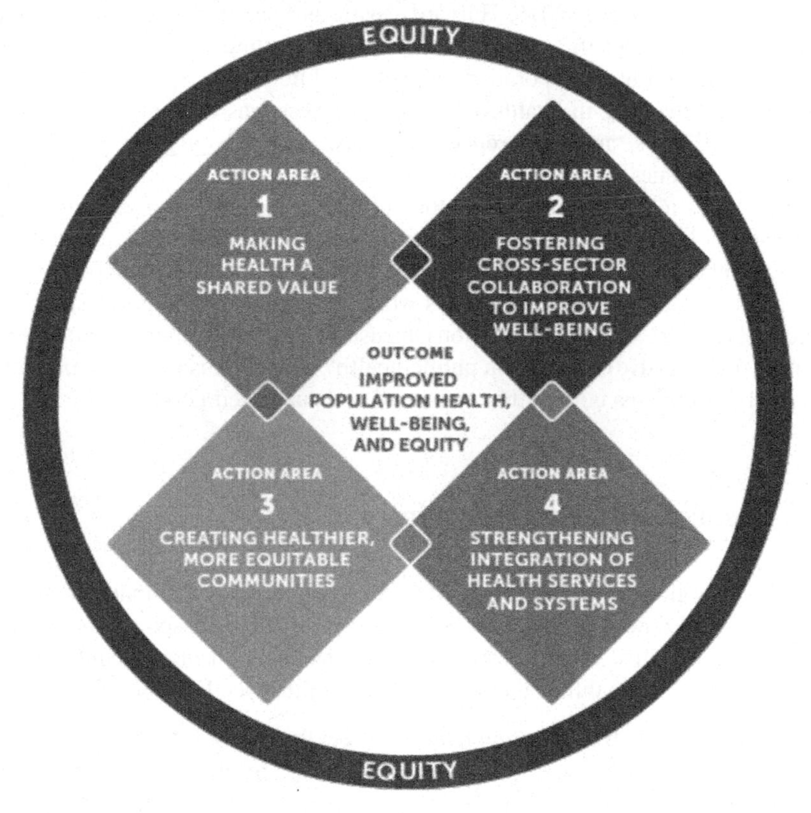

[56] Ibid.

[57] The Robert Wood Johnson Foundation, "From Vision to Action: A Framework and Measures to Mobilize a Culture of Health," March 2016, http://www.rwjf.org/content/dam/COH/RWJ000_COH-Update_CoH_Report_1b.pdf, accessed May, 2016.

Source: The Robert Wood Johnson Foundation, "From Vision to Action: A Framework and Measures to Mobilize a Culture of Health," March 2016, http://www.rwjf.org/content/dam/COH/RWJ000_COH-Update_CoH_Report_1b.pdf, accessed May, 2016. *Copyright 2016. Robert Wood Johnson Foundation. Used with permission from the Robert Wood Johnson Foundation.*

In reference to the reasoning behind the Action Framework, RWJF stated:

> The Action Framework reflects a vision of health and well-being as the sum of many parts, addressing the interdependence of social, economic, physical, environmental, and spiritual factors. It is intended to generate unprecedented collaboration and chart our nation's progress toward building a Culture of Health. Equity and opportunity are overarching themes of the entire Action Framework—not merely to highlight our nation's health disparities, but to move toward achieving health equity. The Action Framework groups the many actors, and the many facets, of a Culture of Health into four Action Areas each connected to and influenced by the others.[58]

And finally, in April, 2016, Harvard Business School, the Harvard T.H. Chan School of Public Health, and RWJF hosted the Culture of Health conference—which focused on how corporations affected public health and what benefits they gained from investing in health. Although progress was evident, these programs, research initiatives, and conferences typically included only the most forward-thinking companies.

Many questions remained about the future of business involvement in public health on a larger scale: How did businesses intentionally and unintentionally impact public health? What benefits could businesses gain from supporting public health? How could business executives work with public health professionals more effectively? How should corporations measure and report their cumulative (i.e., positive and negative) impacts on public health? Could a business benefit itself and society by adopting a culture that embraced and supported good health?

The Population Health Footprint

Every corporation, wittingly or unwittingly through decisions made or not made every day, laid down a population health footprint (PHF) based on its cumulative positive and negative effects across four dimensions: consumer health, employee health, community health, and environmental health. (See Exhibit 1.6):

1. **Consumer Health:** How organizations affect the safety, integrity, and healthfulness of the products and services they offer to their customers and end consumers.
2. **Employee Health:** How organizations affect the health of their employees (e.g., provision of employer-sponsored health insurance, workplace practices and wellness programs).

[58] Ibid.

3. **Community Health:** How organizations affect the health of the communities in which they operate and do business.
4. **Environmental Health:** How organizations' environmental policies (or lack thereof) affect individual and population health.

Exhibit 1.6: Population Health Footprint (PHF)

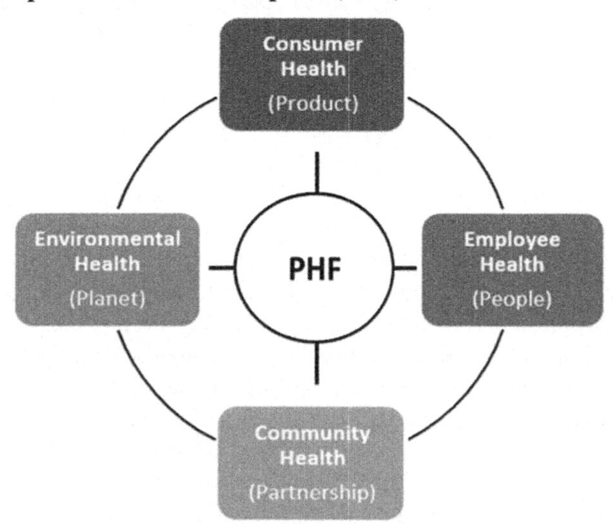

Source: Casewriter, SHINE initiative.

This four-way classification dovetailed with SHINE research, which focused on how businesses affected health through "products, people, and planet." The four-way classification deliberately separated people effects into employee health and community health, which in practice largely involved "partnerships" between the private and public sectors.

Establishing a Corporate Culture of Health

While the Culture of Health vision that RWJF articulated included a role for business, it was an overall vision for society at large. To participate in that vision, companies needed an internal business culture that encouraged and motivated engagement in public health efforts.

Within all companies, decisions are made every day that impact public health across these four areas. A corporation with a corporate Culture of Health was one in which improving public health was articulated by all employees—from the CEO down to front-line workers—as part of the corporation's mission and considered prominently in decision-making at all levels. Improving health was considered a vital part of the corporation's strategy, and corporate impacts on consumer health, employee health, community health, and environmental health were all measured and reported. A Culture of Health was also a culture of science. From designing new products to nudging employee behaviors, evidence and science had to underpin health-related program development and adoption.

However, in 2016, very few companies integrated their health-related efforts, included health as a corporate value, or calculated their total population health footprint. Companies often took a fragmented approach to addressing health concerns. Typically, product design and quality assurance managers attended to the health and safety of products. In most companies, a benefits manager in the human resources department designed and implemented programs to promote employee wellbeing, reduce insurance costs, and boost productivity. Many corporations increasingly pursued innovative wellness programs and advanced technology to measure employee health, such as giving their employees wearable fitness-tracking devices like Fitbits.[59] A vice president for community relations or the head of the company's foundation might allocate some funds to community health projects, while a director of sustainability might address environmental health through reductions in the company's carbon footprint and water use. However, even in corporations that pursued initiatives in all four areas, such efforts remained fragmented across different departments; the whole was no greater than the parts, and any new health initiative was simply bolted onto existing business activities. As a result, very few companies and their chief executives embraced a culture of health as a strategic imperative.

In 2016, the mention of "health" to most corporate leaders brought to mind the cost of health insurance—not how the company could invest to prevent further escalation in societal health care costs (which, ultimately, fell to nearly every individual and every company). Further, even companies in the health care industry, which arguably should have been in the vanguard, did not measure their population health footprints in a comprehensive way. They might include the word "health" in their mission statements, but CEOs rarely stressed health as a core value to be considered every time an employee made a decision.

At its core, adopting a Culture of Health required a shift in how business thought and talked about health—not unlike the change in opinion that brought environmental sustainability into the mainstream a few years ago. At a corporation with this new, health-focused vision, health was not treated as an afterthought or only in the contexts of regulatory compliance or corporate social responsibility—it was a central tenet of doing business. Consider Royal Caribbean, a cruise ship operator, which had many initiatives across all four areas of the PHF. The organization promoted a culture of "Above and Beyond Compliance," continuously setting health, safety, and environmental performance targets that exceeded current regulatory standards, thereby leading the industry towards a healthier new normal.

However, adopting a Culture of Health was a journey. Where a company was on the continuum varied depending on the company's products and services, consumer base, and geographic location, among other things. A corporation that produced a product that was inherently less healthy than others (e.g., soda compared to water)

---

[59] Jason Cipriani, "Here's Why Fitbit is giving Target 335,000 fitness-tracking devices," *Fortune*, September 16, 2015, http://fortune.com/2015/09/16/fitbit-hipaa/, accessed May, 2016.

could still pursue a Culture of Health by developing the necessary competencies to shift to a healthier product portfolio. For example, PepsiCo Inc. took a long-term approach to health, becoming more transparent about its products' nutritional value and creating the Global Nutrition Group, a division that focused on creating healthier options for consumers.[60]

Common Barriers

Most businesses remained reluctant to see health as an investment and an opportunity rather than as an expense. There were several common barriers to creating a Culture of Health. These included:

Perceived conflict with other business objectives: Some corporate leaders reasoned that the promotion of health was at odds with other business goals, notably maximizing value to shareholders. Though researchers had found that companies that prioritized health and safety objectives outperformed the S&P average, there was an ongoing perception that pursuing a culture of health was an expense with no certain return.[61]

In New York, for example, small retailers and restaurants feared losing revenues and profits as a result of regulations requiring nutrition disclosures, calorie counts, and soda serving size information, but there was no evidence that this had happened.[62] The ability to invest in employee health benefits was often associated with higher margin companies like Starbucks. Yet, at Costco Wholesale, a low cost, membership-only retailer, the average employee in 2013 made $20.89/h (when the minimum wage was $7.25 an hour) and 88 % of its employees participated in company-sponsored health insurance.[63] CEO Craig Jelinek said, "I just think people need to make a living wage with health benefits. It also puts more money back into the economy and creates a healthier country. It's really that simple."[64]

---

[60] Rosabeth Kanter, Rakesh Khurana, Rajiv Lal, Eric Baldwin, "PepsiCo, Performance with Purpose, Achieving the Right Global Balance," HBS No. 412-079 (Boston: Harvard Business School Publishing, 2012).

[61] Raymond Fabius MD; Loeppke, Ronald R. MD, MPH; Hohn, Todd CSP; Fabius, Dan DO; Eisenberg, Barry CAE; Konicki, Doris L. MHS; Larson, Paul MS, "Tracking the Market Performance of Companies That Integrate a Culture of Health and Safety: An Assessment of Corporate Health Achievement Award Applicants," *Journal of Occupational and Environmental Medicine,* January 2016, Volume 58, Issue 1, pp. 3–8, accessed at http://journals.lww.com/joem/Abstract/2016/01000/Tracking_the_Market_Performance_of_Companies_That.2.aspx, accessed January, 2016.

[62] Stanford GSB staff, "Researchers: How Does Posting Calories Affect Behavior?," *Stanford Business,* February, 1, 2011, https://www.gsb.stanford.edu/insights/researchers-how-does-posting-calories-affect-behavior, accessed May, 2016.

[63] Brad Stone, "Costco CEO Craig Jelinek Leads the Cheapest, Happiest Company in the World," *Bloomberg, June 7, 2013,* http://www.bloomberg.com/bw/articles/2013-06-06/costco-ceo-craig-jelinek-leads-the-cheapest-happiest-company-in-the-world, accessed April, 2016.

[64] Ibid.

Corporate values overlooked health: Few companies outside of the healthcare sector made health central to their mission. The word "health" did not appear in the mission statements of any Fortune 100 companies (outside of the healthcare sector) in 2016.[65]

This was due to several causes. First, healthcare was often seen as a seemingly intractable political challenge that the government, rather than the private sector, should solve. This meant that corporate involvement was not expected. This was even truer in countries with universal, government-funded health care, such as many in Europe. Compounding the challenge, many viewed health-related lifestyle decisions as a matter of personality. Moreover, some health issues, especially those related to employee health, involved thorny questions of personal privacy that corporations wished to avoid. And lastly, business often prioritized other social issues, like educating children. Businesses intuitively understood that a good education would develop the skills they needed in the next generation of employees. However, there was less understanding around the necessity of healthy communities and the many connections between education and health. Healthy children learned better in school, and educated children were more likely to model healthy behaviors.

Therefore, for all of these reasons, promoting a culture of health required a shift in corporate mission and vision, and corporate leaders always faced significant challenges when advocating changes to core corporate values.[66]

Inadequate measures for health reporting: As of 2016, there were no comprehensive methodologies to measure and compute the cumulative costs and benefits of a company's four-part population health footprint. Some progress has been made in measuring employee health and environmental health, but the lack of reporting standards made it difficult for senior executives to prioritize health initiatives.

Measuring financial and non-financial business activities and outputs was not new; in the 1990s, Robert Kaplan and David Norton made the case for measuring and reporting non-financial in addition to financial metrics as part of their balanced scorecard.[67] Corporate reporting of non-financial metrics continued to evolve, and with the explosion of corporate social responsibility (CSR) programs throughout the 1990s and 2000s, more companies began to report these metrics routinely. By 2013, 76 of the "Fortune 100" companies published social responsibility reports separate from their annual corporate reports.[68]

Specific measurement systems appeared for reporting and ranking sustainable and ethical activities. PwC's Total Impact Measurement and Management moved beyond shareholder value to calculate the social and environmental impacts as well

---

[65] Casewriter analysis.

[66] Harold L. Sirkin, Perry Keenan, and Alan Jackson, "The Hard Side of Change Management," *Harvard Business Review,* October 2005.

[67] Robert S. Kaplan and David P. Norton, "Putting the Balanced Scorecard to Work," *Harvard Business Review,* September 1, 1991.

[68] Rachel A. Spero, Fred D. Ledley, "Making Public Health Central to Standards for Corporate Social Responsibility," *Center for Integration of Science and Industry: Departments of Natural & Applied Science, Management,* Bentley University, 2015.

as the economic and tax impacts of a company's activities.[69] There were numerous global rankings of companies, such as Fortune's 100 Best Companies to Work For and Ethisphere's World's Most Ethical Companies. The Fortune ranking included detailed questions on topics tangential to health—from benefit programs to diversity—but there was no aggregate health component of each company's total score and the questionnaire was not publicly available.[70] The Ethisphere corporate ethics quotient (EQ) covered five topic areas, none of them health-related.[71]

Moreover, CSR reporting rarely included metrics for health.[72] Corporate impacts on public health were measured only as an inferred benefit from environmental performance improvements. Researchers stated:

> Examining a number of proposed standards for CSR, we found that public health concerns are notably absent. We argue that the emergence of accountability standards for CSR represents an opportunity to advance public health that is not being addressed, and ask what steps need to be taken to make public health concerns central to the standards for CSR in the future.[73]

Population health was so entwined with education, employment and other social services that agreeing on how to cleanly measure cause and effect remained challenging.

Modest public concern about health: Corporations were often motivated to pursue social issues based on their importance to consumers; healthcare rarely surfaced among the top issues of concern to the public.[74] A *Wall Street Journal/NBC News* poll found that the percentage of people who thought "health care" was a top issue for the federal government was 13 % in April 2015 and 9 % in December 2015.[75] "National security and terrorism" rose from 21 to 40 % in the same period.[76] In addition, Pew Research Center's findings on the American public's top priorities in 2014 showed that "Securing Medicare" and "Reducing health care costs" came in below issues related to the economy, education, and terrorism (see Table 1.4). Furthermore,

---

[69] PWC, "Total impact Measurement and Management," http://www.pwc.com/gx/en/services/sustainability/publications/total-impact-measurement-management/total.html, accessed April, 2016.

[70] Fortune, "100 Best Companies to Work For," *Fortune website,* http://fortune.com/best-companies/, accessed April, 2016.

[71] Ethisphere, "Scoring and Methodology," *Ethisphere website,* http://worldsmostethicalcompanies.ethisphere.com/scoring-methodology/, accessed April, 2016.

[72] Rachel A. Spero, Fred D. Ledley, "Making Public Health Central to Standards for Corporate Social Responsibility," *Center for Integration of Science and Industry: Departments of Natural & Applied Science, Management,* Bentley University, 2015.

[73] Ibid.

[74] Pew Research Center, "Thirteen Years of the Public's Top Priorities," January 27, 2014, http://www.people-press.org/interactive/top-priorities/, accessed March, 2016.

[75] Janet Hook, "Poll Finds National Security Now a Top Concern," *The Wall Street Journal,* December 14, 2015, http://www.wsj.com/articles/poll-finds-national-security-now-a-top-concern-1450130463, accessed April, 2016.

[76] Ibid.

Table 1.4 Top 10 public priorities in 2014

Issue	Percent of public identifying issue as a "top priority" (%)
Strengthening nation's economy	80
Improving job situation	74
Defending against terrorism	73
Improving education	69
Securing Social Security	66
Reducing budget deficit	63
Securing Medicare	61
Reducing health care costs	59
Reducing crime	55
Reforming the nation's tax system	55

Source: Pew Research Center, "Thirteen Years of the Public's Top Priorities," *Pew Research Center*, Washington, DC, January 27, 2014, http://www.people-press.org/interactive/top-priorities/, accessed March 2016

entrance and exit polls during the 2016 presidential primaries showed that voters considered the economy and terrorism to be top concerns.[77]

This relative lack of concern was likely due to several causes. First, as previously discussed, public health practitioners had achieved significant advances in the field.[78] Average life expectancies had vastly improved.[79] Second, research by NPR, the Robert Wood Johnson Foundation, and the Harvard T.H. Chan School of Public Health indicated that Americans tended to rate their own health care and associated costs more positively than the overall system.[80] Finally, economic development reports for nations, states, and cities frequently emphasized the importance of high quality high school and college education for successful economic and job growth. Despite strong evidence that there were connections between good health, educational achievement and economic growth, health was mentioned far less often.[81]

[77] Emily Swanson, "The top issues influencing voters in South Carolina and Nevada," *PBS Newshour,* February 21, 2016, http://www.pbs.org/newshour/rundown/the-top-issues-influencing-voters-in-south-carolina-and-nevada/, accessed April, 2016.

[78] National Institute on Aging, "Health and Aging: Living Longer," *U.S. Department of Health and Human Services,* 2011, https://www.nia.nih.gov/research/publication/global-health-and-aging/living-longer, accessed April, 2016.

[79] Rosabeth Moss Kanter, Howard Koh, Pamela Yatsko, "The State of U.S. Public Health: Challenges and Trends," HBS No. 316-001 (Boston: Harvard Business School Publishing, 2015).

[80] NPR, Robert Wood Johnson Foundation, Harvard T.H. Chan School of Public Health, "PATIENTS' PERSPECTIVES ON HEALTH CARE IN THE UNITED STATES: A LOOK AT SEVEN STATES & THE NATION," February 2016, accessed at http://www.npr.org/assets/img/2016/02/26/PatientPerspectives.pdf, accessed April, 2016.

[81] World Health Organization (WHO), "Economic growth," *WHO website,* http://www.who.int/trade/glossary/story019/en/, accessed April, 2016.

The Case for Engagement

These barriers were outweighed by the benefits corporations could reap from investing in health. There were four main reasons for businesses to focus on public health: to address a moral imperative; to reduce costs; to increase revenues; and to improve their reputations.

Address moral imperative: The most basic reason for corporations to address their impacts on public health was that there was a moral imperative to do so. First, businesses' products and services largely defined and framed culture in capitalist societies. Second, many businesses caused negative externalities (negative impacts on society for which they did not have to pay). For example, most of the world's population bought food from the private sector, yet only very few American's diets met the government's dietary guidelines.[82] Though correcting the unhealthy consequences of unhealthy food products were externalities that were not charged to food manufacturers, leaders of major companies such as Mars and Nestle recognized increasingly that they had to be part of the solution.

Further, the business community was uniquely well-positioned to enact change—it had the ability to scale solutions, the necessary financing flexibility that governments, non-profits, and non-governmental organizations (NGOs) did not, and it was skilled at managing and leading change. Businesses knew how to use marketing to persuade consumers—skills that could also help persuade individuals and communities to adopt preventive health care behaviors.

Reduce costs: Healthcare costs in aggregate continued to consume a progressively larger portion of the overall US GDP reaching 17.5 % by 2015.[83] Business investments in health could cut unnecessary costs. For example, investments in employee health could reduce losses associated with absenteeism and presenteeism.[84] Employers also focused on stemming the cost of employer-sponsored health insurance premiums; these accounted for 20 % of all health care expenditures in the US and had outpaced inflation since 2000.[85] Moreover, investing in workplace safety and health could reduce fatalities, injuries and illnesses, which created a wide range of cost savings (e.g., lowering workers' compensation costs, avoiding OSHA penal-

[82] Susan M. Krebs-Smith, Patricia M. Guenther, Amy F. Subar, Sharon I. Kirkpatrick, and Kevin W. Dodd, "Americans Do Not Meet Federal Dietary Recommendations," *Journal of Nutrition*, 2010 Oct; 140(10): 1832–1838, accessed at http://www.ncbi.nlm.nih.gov/pmc/articles/PMC2937576/, accessed April, 2016.

[83] The World Bank, "Data: Health expenditure, total (% of GDP)." *The World Bank website,* http://data.worldbank.org/indicator/SH.XPD.TOTL.ZS, accessed April, 2016.

[84] Thomas Parry and Bruce Sherman, "Workforce Health—The Transition from Costs to Outcomes to Business Performance," *Benefits Quarterly first quarter 2015,* http://www.ifebp.org/inforequest/ifebp/0166489.pdf, accessed October, 2015.

[85] The Kaiser Family Foundation and Health Research & Educational Trust, "Employer Health Benefits: 2015 Summary Findings," September 22, 2015, http://kff.org/health-costs/report/2015-employer-health-benefits-survey/, accessed October, 2015.

ties, and reducing the costs associated with hiring and training replacement employees).[86] Corporate support of community health initiatives also had the potential to reduce employee health costs, as the behaviors of employees and their families were shaped by attitudes and behaviors prevalent in the local communities where they lived.[87]

Some corporations also faced significant costs due to poor attention to consumer health. For example, a product recall would cost a company the out-of-pocket costs of managing the recall operation, lost revenues from the recalled products it could no longer sell, and a damaged brand reputation.[88] In the area of environmental health, corporations could achieve cost savings as a result of improved water and energy efficiency.[89]

Increase revenues: Businesses increasingly realized that the promotion of health facilitated broader economic growth, and a strong economy was invariably good for top-line revenues.[90] Research had shown that "the relationship between health and economy runs both ways, lasts throughout an individual's lifetime and is intergenerational."[91]

In addition to the potential for broad economic growth, there were specific opportunities for corporations to increase revenue by considering health impacts. Within consumer health, research indicated that a majority of consumers—especially younger consumers—were willing to pay a premium for socially responsible brands.[92] Companies invested in their communities were more likely to retain the loyalty of their people in a recession. Investing in employee health had the potential to promote employee satisfaction, engagement, and therefore retention.[93] Further, employees were often not just interested in what their employers did within the

---

[86] Occupational Safety & Health Administration (OSHA), "Business Case for Safety and Health," *OSHA website,* https://www.osha.gov/dcsp/products/topics/businesscase/, accessed April, 2016.

[87] Vera Oziransky, Derek Yach, Tsu-Yu Tsao, Alexandra Luterek, Denise Stevens, "Beyond the Four Walls: Why Community Is Critical to Workforce Health," The Vitality Institute, July 2015, accessed at http://thevitalityinstitute.org/site/wp-content/uploads/2015/07/VitalityInstitute-BeyondTheFourWalls-Report-28July2015.pdf, accessed November, 2015.

[88] Kathleen Cleeran, "Using advertising and price to mitigate losses in a product-harm crisis," *Kelley School of Business, Indiana University,* 2015, 58, 157–162, p. 160.

[89] McKinsey & Company, "The business of sustainability: McKinsey Global Survey results," McKinsey & Company website, October, 2011, accessed at http://www.mckinsey.com/insights/energy_resources_materials/the_business_of_sustainability_mckinsey_global_survey_results, accessed January, 2016.

[90] Rifat Atun, Claire Chaumont, Joseph R Fitchett, Annie Haakenstad, Donald Kaberuka, "Poverty Alleviation and the Economic Benefits of Investing in Health," *Forum for Finance Ministers 2016*, accessed at https://cdn2.sph.harvard.edu/wp-content/uploads/sites/61/2015/09/L-MLIH_Health-economic-growth-and-development_Atun-and-Kaberuka_4-11-16.pdf, accessed May, 2016.

[91] Ibid.

[92] Nielsen, "CONSUMER-GOODS' BRANDS THAT DEMONSTRATE COMMITMENT TO SUSTAINABILITY OUTPERFORM THOSE THAT DON'T," *Nielsen Press Room,* October 12, 2015, http://www.nielsen.com/us/en/press-room/2015/consumer-goods-brands-that-demonstrate-commitment-to-sustainability-outperform.html, accessed April, 2016.

[93] Zeynep Ton, "The Good Jobs Strategy," *Houghton Mifflin Harcourt Publishing Company*, New York, New York, 2014, pp. 64–67.

arena of employee health; they were attracted and motivated by what the company did across all four pillars.

And evidence suggested that these kinds of investments paid off. In 2016, researchers assessed the stock market performance of companies that achieved high scores on either health or safety in the Corporate Health Achievement Award (CHAA)[94] process and found that the share prices of companies that prioritized health and safety outperformed the S&P average on all tests.[95] Further, a 2012 analysis that examined corporations that had adopted sustainability policies[96] by 1993 found that those firms were more likely to have established processes for stakeholder engagement, to be more long-term oriented, and to exhibit higher measurement and disclosure of nonfinancial information over time.[97] The researchers also found that these companies' share prices significantly outperformed their counterparts over the long-term.[98]

Improve reputations: Corporations could also improve their reputations by promoting health. Results from Edelman's 2016 Trust Barometer showed that trust in business was low.[99] There was ample opportunity for business, individually and collectively, to improve consumer perceptions and relationships by helping to solve important societal challenges.[100]

Table 1.5 matrix arranges the four public health pillars against these business objectives. The check marks indicate the objectives which were typically associated with each health impact area. However, the relative importance of each pillar varied by company or industry. For example, asset-heavy industrial companies could contribute greatly to public health by reducing their environmental health footprints. Retailers on the other hand, might contribute most by assuring the healthfulness of the products in their stores and by investing in the health of the communities from which they drew their consumers and employees.

---

[94] The CHAA was established in 1995 by the American College of Occupational and Environmental Medicine (ACOEM) to recognize the healthiest, safest companies and organizations in North America and to raise awareness of best practices in workplace health and safety programs.

[95] Raymond Fabius MD; Loeppke, Ronald R. MD, MPH; Hohn, Todd CSP; Fabius, Dan DO; Eisenberg, Barry CAE; Konicki, Doris L. MHS; Larson, Paul MS, "Tracking the Market Performance of Companies That Integrate a Culture of Health and Safety: An Assessment of Corporate Health Achievement Award Applicants," *Journal of Occupational and Environmental Medicine,* January 2016, Volume 58, Issue 1, pp. 3–8, accessed at http://journals.lww.com/joem/Abstract/2016/01000/Tracking_the_Market_Performance_of_Companies_That.2.aspx, accessed January, 2016.

[96] Corporate policies related to the environment, employees, community, products, and customers.

[97] Robert G. Eccles, Ioannis Ioannou, George Serafeim, "The Impact of Corporate Sustainability on Organizational Processes and Performance," *The National Bureau of Economic Research,* March 2012, http://www.nber.org/papers/w17950, accessed April, 2016.

[98] Ibid.

[99] Edelman, "2016 Executive Summary: Edelman trustbarometer," *Edelman website,* http://www.edelman.com/insights/intellectual-property/2016-edelman-trust-barometer/executive-summary/, accessed May, 2016.

[100] Ibid.

Table 1.5 Business objectives associated with PHF components

	Consumer health	Employee health	Community health	Environmental health
Address moral imperative	X	X	X	X
Reduce costs	X	X		X
Increase revenues	X	X	X	X
Improve reputations			X	X

Source: Casewriter analysis

Few companies analyzed their total health footprints across the four areas. As a result, investments in improving a company's PHF were not allocated optimally. Most companies did not consciously allocate resources to initiatives across the four areas such that the public health return on each dollar invested was the same. Some corporations focused their effort almost entirely within one area, overlooking the potential benefits that might come from investing some of their resources in the others.

How to Engage: Creating Shared Value

Traditionally, corporations engaged with societal challenges through corporate social responsibility (CSR) programs, which could both make a positive social impact and enhance brand differentiation.[101] There was also increasing evidence that companies that invested in CSR programs performed better financially (though some believed it was better financial performance that allowed a company the luxury of directing some portion of profits to CSR).[102]

However, as John Browne discussed in his book *Connect,* corporate efforts had to go beyond traditional CSR.[103] While philanthropy could be impactful and worthy, creating lasting change and improved corporate reputation required businesses to incorporate health impact considerations into their business strategies. In an era of growing transparency where the broader stakeholder impacts of business actions were increasingly evaluated, social and business strategies became interlinked.

[101] James Epstein-Reeves, "Six Reasons Companies Should Embrace CSR," *Forbes,* February 21, 2012, accessed at http://www.forbes.com/sites/csr/2012/02/21/six-reasons-companies-should-embrace-csr/, accessed November, 2015.

[102] Rachel A. Spero, Fred D. Ledley, "Making Public Health Central to Standards for Corporate Social Responsibility," *Center for Integration of Science and Industry: Departments of Natural & Applied Science, Management,* Bentley University, 2015.

[103] John Browne with Robin Nuttal and Tommy Stadlen, "Connect: How Companies Succeed by Engaging Radically with Society," Virgin Digital, September 10, 2015.

In 2006, Michael Porter and Mark Kramer introduced the idea of shared value—that corporations could connect societal progress and business returns to advance mutually shared interests. In a 2011 *Harvard Business Review* article, they wrote:

> The concept of shared value can be defined as policies and operating practices that enhance the competitiveness of the company while simultaneously advancing the economic and social conditions in the communities in which it operates. Shared value creation focuses on identifying and expanding the connections between societal and economic progress.[104]

Given the wide range of benefits corporations could reap by considering health impacts, the concept of shared value could be applied easily to health promotion efforts.[105] For example, in 2009, Intel Corporation partnered with local healthcare providers in Portland, Oregon, and using its expertise in supply chain management to improve treatment paradigms for conditions such as diabetes and lower back pain.[106] This not only improved care and reduced healthcare costs for Intel employees, but it also improved care for all people in the community using the participating hospitals.[107]

Conclusion

In sum, all corporations laid down a population health footprint—intentionally or unintentionally—through their positive and negative contributions in the areas of consumer health, employee health, community health, and environmental health. By 2016, few corporations were investing systematically in all four areas. Rather, corporate contribution to advancing public health remained largely fragmented.

However, some corporations were beginning to adopt a Culture of Health that prioritized health and encouraged all employees, from entry-level workers to CEOs, to actively consider the health effects of everyday business decisions. These corporations realized that embracing a Culture of Health not only contributed to meaningful societal change, but also returned significant business benefits. Despite such changes within a handful of forward-thinking organizations, broader corporate prioritization of health was necessary. Further research was necessary on how to measure and validate a company's population health footprint before large numbers of corporations were likely to come on board.

---

[104] Michael E. Porter and Mark R. Kramer, "Creating Shared Value: How to reinvest capitalism—and unleash a wave of innovation and growth," *Harvard Business Review January-February 2011*, p. 6.

[105] Ibid.

[106] Patricia A. McDonald, Robert S. Mecklenburg, and Lindsay A. Martin, "The Employer-Led Health Care Revolution," *Harvard Business Review July–August 2015*.

[107] Ibid.

Chapter 2
Consumer Health

In 2012, New York City Mayor Michael Bloomberg proposed a ban on sugary dinks (e.g., sodas, teas, and energy drinks) larger than 16 oz. in restaurants, delis, sports arenas, movie theaters and food carts.[1] The bold public health decision was made in an effort to curb obesity and diabetes rates in the city, where more than half of the adult population was overweight or obese.[2] While some lauded the measure for its health-promoting intentions, many others believed that the policy would harm small business profits and too severely limited the freedom of consumers to make their own choices.[3] Ultimately, the state's highest court overturned the measure.[4] Despite the failure of the ban in New York City, the proposition highlighted an increasing focus on the role of foods and beverages—and sugary drinks in particular—in the obesity epidemic.

Coupled with regulatory threats, food and beverage companies also increasingly faced more health-conscious consumers, as well as public health campaigns on the consequences of high-calorie food and drink consumption—and it was clear these

---

Reprinted with permission of Harvard Business School Publishing
Consumer Health, HBS No. 9-516-076
This case was prepared by John A. Quelch and Emily C. Boudreau.

[1] Alice Park, "The New York City Soda Ban, and a Brief History of Bloomberg's Nudges," *TIME*, May 31, 2012, http://healthland.time.com/2012/05/31/bloombergs-soda-ban-and-other-sweeping-health-measures-in-new-york-city/, accessed December 2015.

[2] MICHAEL M. GRYNBAUM, "Health Panel Approves Restriction on Sale of Large Sugary Drinks," *The New York Times,* September 13, 2012, http://www.nytimes.com/2012/09/14/nyregion/health-board-approves-bloombergs-soda-ban.html, accessed December 2015.

[3] Ibid.

[4] MICHAEL M. GRYNBAUM, "New York's Ban on Big Sodas Is Rejected by Final Court," *The New York Times,* June 26, 2014, http://www.nytimes.com/2014/06/27/nyregion/city-loses-final-appeal-on-limiting-sales-of-large-sodas.html, accessed December 2015.

J.A. Quelch, E.C. Boudreau, *Building a Culture of Health*, SpringerBriefs in Public Health, DOI 10.1007/978-3-319-43723-1_2

combined forces were taking effect.[5] In a study published in 2014, researchers noted that overall caloric intake had declined between 2003 and 2011.[6] What's more, they discovered that the decline was strongest in children and was disproportionately due to changes in beverage consumption rather than food.[7] Interestingly, the overall liquid refreshment beverage market had grown between 2010 and 2015, due to growth in bottled water, sports drinks, energy drinks, and teas and coffees.[8] By 2015, sales of full-calorie soda in the US had gone down by more than 25 % over the previous 20 years.[9] (See Exhibit 2.1 for US bottled water and soda consumption per capita over time.)

Exhibit 2.1: US Soda and Bottled Water per Capita Consumption (Gallons)

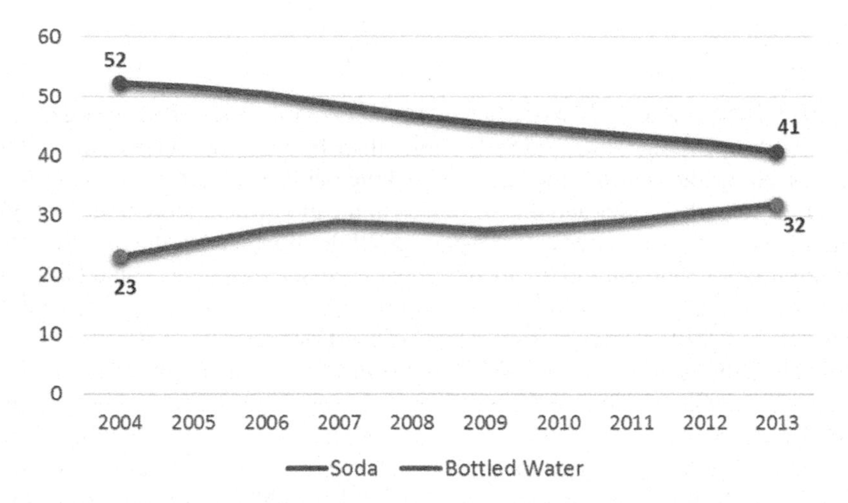

Source: Casewriter's diagram based on market data from "The 2015 Beverage Marketing Directory," Marketing Beverage Corporation, pp. xxi–xxv, accessed February 2016.

Large soft drink companies responded in different ways to these market challenges. In 2015, many public health experts accused The Coca-Cola Company

[5] Rosabeth Kanter, Rakesh Khurana, Rajiv Lal, Eric Baldwin, "PepsiCo, Performance with Purpose, Achieving the Right Global Balance," HBS No. 412-079 (Boston: Harvard Business School Publishing, 2012).

[6] Shu Wen Ng, Meghan M Slining, and Barry M Popkin, "Turning point for US diets? Recessionary effects or behavioral shifts in foods purchased and consumed," *The American Journal of Clinical Nutrition,* 2014; 99:609–16.

[7] Ibid.

[8] Trefis Team Contributor, "Sports Drink Wars In The U.S. To Get Exciting?," *Forbes,* August 18, 2015, accessed at http://www.forbes.com/sites/greatspeculations/2015/08/18/sports-drink-wars-in-the-u-s-to-get-exciting/#4b29ee257c9f, accessed January 2016.

[9] Margot Sanger-Katz, "The Decline of 'Big Soda'," *The New York Times,* October, 2, 2015, accessed at http://www.nytimes.com/2015/10/04/upshot/soda-industry-struggles-as-consumer-tastes-change.html, accessed December 2015.

("Coke") of funding misleading public health research to disguise the role of soft drinks in the obesity epidemic. In particular, a non-profit group called the Global Energy Balance Network (GEBN), which placed a research emphasis on the exercise-related causes of obesity rather than dietary ones, came under fire for accepting over $1 million in funding from Coke.[10] Following several media articles on the relationship between the nonprofit and Coke, Coke's chief science and health officer stepped down and GEBN shut down due to lack of funding.[11,12] Muhtar Kent, CEO of stated in the *Wall Street Journal*, "At Coca-Cola, the way we have engaged the public health and scientific communities to tackle the global obesity epidemic that is plaguing our children, our families and our communities is not working."[13]

In comparison, PepsiCo Inc. ("Pepsi") took a longer-term approach, becoming more transparent about its products and seeking new, healthier areas of business. In 2010, it created the Global Nutrition Group, a division of the company which focused on broadening its product portfolio to create healthier options for consumers.[14] In December 2015, Pepsi announced that it would launch a new food and beverage vending machine called "Hello Goodness" in 2016.[15] Pepsi stated that "the vending unit offers good- and better-for-you product choices from the company's diverse and highly complementary food and beverage portfolio."[16] (See Exhibit 2.2 for the "Hello Goodness" vending machine.)

[10] Nancy Fink Huehnergarth, "Emails Reveal How Coca-Cola Shaped The Anti-Obesity Global Energy Balance Network," *Forbes,* November 24, 2015, http://www.forbes.com/sites/nancyhuehnergarth/2015/11/24/emails-reveal-how-coca-cola-shaped-the-anti-obesity-global-energy-balance-network/, accessed December 2015.

[11] ANAHAD O'CONNOR, "Coke's Chief Scientist, Who Orchestrated Obesity Research, Is Leaving," *The New York Times,* November 24, 2015, http://well.blogs.nytimes.com/2015/11/24/cokes-chief-scientist-who-orchestrated-obesity-research-is-leaving/, accessed December 2015.

[12] ANAHAD O'CONNOR, "Research Group Funded by Coca-Cola to Disband," *The New York Times,* December 1, 2015, http://well.blogs.nytimes.com/2015/12/01/research-group-funded-by-coca-cola-to-disband/, accessed December 2015.

[13] MUHTAR KENT, "Coca-Cola: We'll Do Better," *The Wall Street Journal,* August 19, 2015, http://www.wsj.com/articles/coca-cola-well-do-better-1440024365?cb=logged0.41843579849228263, accessed December 2015.

[14] Rosabeth Kanter, Rakesh Khurana, Rajiv Lal, Eric Baldwin, "PepsiCo, Performance with Purpose, Achieving the Right Global Balance," HBS No. 412-079 (Boston: Harvard Business School Publishing, 2012).

[15] PepsiCo, "PepsiCo's Hello Goodness™ Vending Initiative Offers Diverse Selection Of Good-And Better-For-You Food And Beverage Products On-The-Go," *PepsiCo website,* December, 14, 2015, http://www.pepsico.com/live/pressrelease/pepsicos-hello-goodness-vending-initiative-offers-diverse-selection-of-goodDOUBLEHYPHENand12142015, accessed January 2016.

[16] Ibid.

Exhibit 2.2: PepsiCo's Hello Goodness™ Vending Machine

Source: PepsiCo, "PepsiCo's Hello Goodness™ Vending Initiative Offers Diverse Selection Of Good-And Better-For-You Food And Beverage Products On-The-Go," *PepsiCo website,* December 14, 2015, http://www.pepsico.com/live/pressrelease/pepsicos-hello-goodness-vending-initiative-offers-diverse-selection-of-good--and12142015, accessed January 2016.

Pepsi viewed these types of decisions as not only responses to growing public health concerns, but also growth-creating opportunities.[17] Many pointed to Indra Nooyi, Pepsi's CEO, as the driving force behind such resolutions. A 2015 *Fortune* article on Nooyi stated:

> From the start of her tenure she dared to acknowledge what was obvious to everyone outside the business but unutterable to those inside it: Junk food makes people fat and harms their health. Nooyi began emphasizing products that are at least a bit healthier than the traditional chips and soda—a pivot some observers thought could sink the company. Now shoppers are proving her right.[18]

---

[17] Rosabeth Kanter, Rakesh Khurana, Rajiv Lal, Eric Baldwin, "PepsiCo, Performance with Purpose, Achieving the Right Global Balance," HBS No. 412-079 (Boston: Harvard Business School Publishing, 2012).

[18] Jennifer Reingold, "PepsiCo's CEO was right. Now what?," *FORTUNE,* June 5, 2015, http://fortune.com/2015/06/05/pepsico-ceo-indra-nooyi/, accessed December 2015.

Despite Pepsi's efforts to move in a new direction, a quarter of its net revenue remained in soft drinks.[19] The corporation was not immune to criticism either. Many public health professionals condemned the entire industry—and especially its largest players—for thwarting public policy efforts aimed at discouraging consumers from drinking soda.[20] Some likened soft drink companies to tobacco companies in the past.[21]

Consumer health meant more than product safety. It included the healthfulness of the products and services the company sold. Cigarette manufacturers complied with product safety standards in the production, manufacturing, and sale of their products; however, these products were still well-documented to be health-harming.[22]

The soda industry's experiences highlighted the consumer health challenges that many kinds of corporations faced by 2015. Consumers were demanding greater transparency and healthier products, and, based on inherent qualities of their product or service, some corporations were more vulnerable to consumer health issues. No one expected all corporations to suddenly change their core business offering or strategy; however, as the juxtaposition of Coke and Pepsi highlighted, some corporations developed greater competencies to address consumer health challenges, even in difficult markets.

Further, the examples raised important questions about the future of consumer health. What role should corporations play in promoting health and wellness through their products and services? How should a corporation respond to providing a product that is less healthy than other available options? What responsibility does the consumer have in choosing to use or buy a product or service? What role should schools of public health and medical schools play in educating the public and future medical leaders on the benefits and risks associated with consumer health? How should the government regulate consumer products and services? Should regulation be primarily driven by local or federal authorities? And finally, what role does the government play in setting and enforcing consumer health standards?

---

[19] Margot Sanger-Katz, "The Decline of 'Big Soda'," *The New York Times,* October, 2, 2015, accessed at http://www.nytimes.com/2015/10/04/upshot/soda-industry-struggles-as-consumer-tastes-change.html, accessed December 2015.

[20] Nancy Gagliardi, "Approval Of Soda Tax In Berkeley Is Scary Precedent For Food Industry," *Forbes,* November 5, 2014, http://www.forbes.com/sites/nancygagliardi/2014/11/05/berkeley-makes-history-on-the-soda-tax/#5c02b0e52e10, accessed February 2016.

[21] Margot Sanger-Katz, "The Decline of 'Big Soda'," *The New York Times,* October, 2, 2015, accessed at http://www.nytimes.com/2015/10/04/upshot/soda-industry-struggles-as-consumer-tastes-change.html, accessed December 2015.

[22] Centers for Disease Control and Prevention (CDC), "Smoking & Tobacco Use" Health Effects of Cigarette Smoking," *CDC website,* http://www.cdc.gov/tobacco/data_statistics/fact_sheets/health_effects/effects_cig_smoking/, accessed December 2015.

Why Corporations Advance Consumer Health

There were many drivers behind corporate investments in consumer health. These included:

Complying with regulation: Consumer safety was a highly regulated arena in many countries. In the US there were several government organizations that monitored the risks of different products and services to consumers, and pursued action against corporations that presented consumer risk (e.g., The Consumer Product Safety Commission, The Federal Trade Commission's Bureau of Consumer Protection, the National Highway Traffic Safety Administration, Bureau of Alcohol, Tobacco, Firearms, and Explosives, the Food and Drug Administration, The Federal Aviation Administration).

Responding to consumer advocacy groups: Consumer advocacy groups were both large and small-scale organizations that worked to protect consumer interests; they established product safety, enforced consumer rights, educated consumers about product or service risks, and compared products[23] (e.g., The Center for Science in the Public Interest, which stated that it "has long sought to educate the public, advocate government policies that are consistent with scientific evidence on health and environmental issues, and counter industry's powerful influence on public opinion and public policies."[24]).

Protecting consumer trust: Brand trust was one of many factors that influenced equity, sales and market share.[25] Consumer trust and confidence were measured globally, as well as by country, industry, and individual company or brand; The Conference Board, Nielsen, Edelman, and the University of Michigan all measured consumer confidence or trust in some capacity.[26,27,28] Edelman's research on consumer trust suggested that corporations often built and maintained trust through integrity, engagement, products and services, and purpose and operations.[29]

---

[23] The Suit Staff, "Understanding Consumer Advocacy Groups," *The Suit Magazine,* May 21, 2015, http://www.thesuitmagazine.com/law-politics/government/22583-understanding-consumer-advocacy-groups.html, accessed, December 2015.

[24] Center for Science in the Public Interest (CPSI), "About CSPI," *CPSI website*, http://www.cspinet.org/about/index.html, accessed December 2015.

[25] Nielsen, "TOP 10 TRUSTED BRANDS: WHAT BRANDS DO MALE AND FEMALE CONSUMERS TRUST THE MOST?," *Nielsen: Newswire website,* 08-31-2015 http://www.nielsen.com/us/en/insights/news/2015/top-10-trusted-brands-what-brands-do-male-and-female-consumers-trust-the-most.html, accessed December 2015.

[26] The Conference Board, "Consumer Confidence Survey," *The Conference Board website* December 29, 2015, https://www.conference-board.org/data/consumerconfidence.cfm, accessed December 2015.

[27] Edelman, "2015 Edelman Trust Barometer," *Edelman website,* http://www.edelman.com/insights/intellectual-property/2015-edelman-trust-barometer/, accessed December 2015.

[28] University of Michigan, "Surveys of Consumers: University of Michigan," *University of Michigan website,* http://www.sca.isr.umich.edu/, accessed December 2015.

[29] Edelman, "Building Trust," *Edelman website,* http://www.edelman.com/insights/intellectual-property/2015-edelman-trust-barometer/building-trust/, accessed December 2015.

Avoiding costs and lost revenue: Corporations that produced and sold harmful products to consumers were at risk for incurring significant profit losses. During a recall, businesses faced the costs of managing a recall operation, the lost revenues from the recalled products that they could no longer sell, and weakened brand equity.[30] Sales often declined when consumers deemed a product or service as risky.[31] Businesses were also liable for consumer harm and could face litigation costs associated with defective products.[32]

Improving social responsibility: Corporate support of consumer safety and health was also part of broader efforts to become more socially responsible. Corporations pursued social responsibility for a number of reasons, including out of moral obligation, for brand differentiation, and for consumer engagement.[33]

Why Corporations Subordinate Consumer Health

Consumer health was not always a top business priority for many reasons; these included:

Conflict with other business objectives: Oftentimes, promoting consumer health came into conflict with other business goals such as creating products that filled consumers' needs and providing value to shareholders. Barring those in the healthcare industry (e.g., pharmaceuticals, wearable fitness tracking technology), products and services in other industries were primarily created for consumer needs unrelated to health. This meant that consumer health and safety was not always a primary objective during design and production.

Furthermore, promoting consumer health could conflict with providing value to shareholders. In some industries, adding safety features might raise unit production costs, lowering unit profit margins. Launching new, healthier products that might cannibalize existing, successful products. Achieving a balance was particularly challenging in the food industry, where one growing segment of consumers called for more natural ingredients, while others criticized the recipe changes that modified their favorite comfort foods.[34]

---

[30] Kathleen Cleeran, "Using advertising and price to mitigate losses in a product-harm crisis," *Kelley School of Business, Indiana University,* 2015, 58, 157–162, p. 160.

[31] Steven Shavell and A. Mitch Polinsky, "The Uneasy Case for Product Liability" Harvard Law School John M. Olin Center for Law, Economics and Business Discussion Paper Series, 2009, Paper 628, p. 1444, accessed at http://lsr.nellco.org/harvard_olin/628/, accessed December 2015.

[32] Steven Shavell and A. Mitch Polinsky, "The Uneasy Case for Product Liability" Harvard Law School John M. Olin Center for Law, Economics and Business Discussion Paper Series, 2009, Paper 628, p. 1444, accessed at http://lsr.nellco.org/harvard_olin/628/, accessed December 2015.

[33] James Epstein-Reeves, "Six Reasons Companies Should Embrace CSR," *Forbes,* February 21, 2012, accessed at http://www.forbes.com/sites/csr/2012/02/21/six-reasons-companies-should-embrace-csr/, accessed November 2015.

[34] Scheherazade Daneshkhu, "Big Food in Health Drive to keep Market Share," *Financial Times,* April, 26, 2016, http://www.ft.com/cms/s/0/83f05ea8-08a3-11e6-a623-b84d06a39ec2.html, accessed May 2016.

In industries such as tobacco and sugar sweetened beverages, advancing consumer health could conflict with providing shareholder value. In some cases, there was a perception that healthier products were or should be more costly, and that only a minority of consumers-even if educated—would make that trade off. In general, younger, educated consumers were more willing to pay a price premium for healthier products and services.[35]

Lack of regulation and/or enforcement: Regulation and its enforcement were less rigorous in certain locations and industries. Consumers International, an independent global campaigning voice for consumers, had independent consumer organizations from across the globe as members.[36] It stated in its 2014 annual survey of member organizations that a majority of its members "viewed existing legislation as ineffective in addressing the key causes of consumer detriment they had identified."[37]

Issues in Consumer Health

Regulation and Innovation

Regulations related to consumer safety, as well as the regulatory bodies that existed to enforce them, were created to protect the public from undue risks. In 2015, the US Consumer Product Safety Commission (CPSC) stated its purpose as:

> CPSC is charged with protecting the public from unreasonable risks of injury or death associated with the use of the thousands of types of consumer products under the agency's jurisdiction…CPSC is committed to protecting consumers and families from products that pose a fire, electrical, chemical, or mechanical hazard.[38]

CPSC had contributed to a reduction in the number of consumer product-related accidents and deaths since its founding.[39] However, regulation also had the potential to hamper innovation that might support consumer health. This was an ongoing issue in the automotive industry, where, by 2015, many manufacturers had designed new technological systems to assist drivers and advance safety. These included

---

[35] Nielsen, "WE ARE WHAT WE EAT HEALTHY EATING TRENDS AROUND THE WORLD," *Nielsen Global Health and Wellness Report—January 2015,* http://www.nielsen.com/content/dam/nielsenglobal/eu/nielseninsights/pdfs/Nielsen%20Global%20Health%20and%20Wellness%20Report%20-%20January%202015.pdf, accessed May 2016.

[36] Consumers International, "Criteria: Becoming a CI Member," *Consumers International website,* http://www.consumersinternational.org/our-members/join-us!/criteria/, accessed December 2015.

[37] Consumers International, "State of Consumer Protection Survey: Summary," 2014–2015, accessed at http://www.consumersinternational.org/media/1568496/ci-survey-summary-2015-english.pdf, accessed December 2015.

[38] United States Consumer Product Safety Commission, "About CPSC," *CPSC website,* http://www.cpsc.gov/en/About-CPSC/, accessed December 2015.

[39] Ibid.

features to detect other cars nearby, brake automatically to avoid pedestrians, and make other automatic movements, such as changing lanes.[40] While most of these safety features aided drivers of manually-driven cars, the potential for driverless cars was on the horizon—representing a paradigm shift in personal vehicle ownership and operation.

Automakers like Audi and Mercedes-Benz were active in this space, but Google led the way from a software intellectual property and hardware perspective.[41] These new types of vehicles created questions for regulators. In December 2015, the California Department of Motor Vehicles announced draft regulations requiring all driverless cars to have a steering wheel, pedals, and a human driver.[42] However, some of the driverless cars Google had been testing in California did not have any manual controls (e.g., steering wheels or pedals).[43] Google spokesman Johnny Luu said:

> In developing vehicles that can take anyone from A to B at the push of a button, we're hoping to transform mobility for millions of people, whether by reducing the 94 % of accidents caused by human error or bringing everyday destinations within reach of those who might otherwise be excluded by their inability to drive a car. Safety is our highest priority and primary motivator as we do this. We're gravely disappointed that California is already writing a ceiling on the potential for fully self-driving cars to help all of us who live here.[44]

While driverless vehicles carried potential benefits for consumer health, there were also concerns about unintended consequences. Because the vehicles would likely be programmed to recognize a pedestrians and avoid hitting them, traffic congestion might increase in busy walking areas, resulting in increased emissions.[45] Further, because cars would have the ability to travel without a passenger, the total number of rides might increase.[46] One commentator wrote:

---

[40] Jessica Guynn and Marco della Cava, "Google 'disappointed' by proposed restrictions on driverless cars," *USA Today,* December 17, 2015, http://www.usatoday.com/story/tech/news/2015/12/16/google-disappointed-by-proposed-rules-from-california-dmv/77447672/, accessed December 2015.

[41] Zachary Hamed, "12 Stocks to Buy If You Believe in Driverless Cars," *Forbes,* January 21, 2015, http://www.forbes.com/sites/zacharyhamed/2015/01/21/driverless-stocks/, accessed December 2015.

[42] Jessica Guynn and Marco della Cava, "Google 'disappointed' by proposed restrictions on driverless cars," *USA Today,* December 17, 2015, http://www.usatoday.com/story/tech/news/2015/12/16/google-disappointed-by-proposed-rules-from-california-dmv/77447672/, accessed December 2015.

[43] BBC News, "Driverless car rules perplexing, says Google," *BBC News,* December 18, 2015, http://www.bbc.com/news/technology-35131538, accessed December 2015.

[44] Jessica Guynn and Marco della Cava, "Google 'disappointed' by proposed restrictions on driverless cars," *USA Today,* December 17, 2015, http://www.usatoday.com/story/tech/news/2015/12/16/google-disappointed-by-proposed-rules-from-california-dmv/77447672/, accessed December 2015.

[45] Drake Bennett, "David Bello: Driverless Tech Could Mean Faster Auto Races," *Bloomberg Businessweek,* April 11, 2016, http://www.bloomberg.com/features/2016-design/a/david-belo/, accessed May 2016.

[46] Ibid.

> Well, if I'm in my driverless vehicle and I can't find a parking space, I could jump out and just get it to circulate until I'm ready for it. Which would add to congestion. Or I could drive to work in the morning, send the car home, and get it to come back for me in the evening, and then go back in it, so then you've got four journeys instead of two.[47]

Driverless automobiles highlighted the challenges public sector regulators faced as a result of rapidly evolving technology. It was often difficult for the regulators to keep pace with innovation. If they cracked down in an overly heavy-handed fashion, they were not only potentially blocking the very entrepreneurial initiative that was America's competitive advantage, but also potentially halting advancements in consumer health. However, there were genuine concerns that, in some cases, new technologies could also negatively affect consumer health.

Regulatory Responsibility

In some industries, businesses exerted a greater level of self-regulation, while in others, government provided substantial oversight. Corporate self-regulation was sometimes deemed a success, such as in the forestry industry, and other times a failure, as was the case in the tobacco industry.[48]

A 2015 incident involving Volkswagen ("VW"), a German automobile manufacturer, showcased why government-led, regulatory oversight was often necessary. VW installed defeat devices in its diesel car engines that were able to detect when regulators were testing the cars, thus enabling the engines to show a lower emissions result in a controlled, testing situation; when the cars were used regularly by consumers, the engines emitted nitrogen oxide pollutants up to 40 times above US standards.[49] Though the investigation began in the US, investigations were opened across the world in countries such as the UK, Italy, France, South Korea, Canada and Germany.[50] The scandal raised consumer health concerns, as environmental conditions were known to affect human health.[51]

Regulatory matters could still become complicated in industries where the burden for regulation and enforcement fell squarely to government actors, as there were often multiple levels of regulators and regulatory enforcers (e.g., international, federal, state, and local organizations) and they could conflict. Such was the case with

[47] Ibid.

[48] Lisa L. Sharma, MBA, MPH, Stephen P. Teret, JD, MPH, and Kelly D. Brownell, PhD, "The Food Industry and Self-Regulation: Standards to Promote Success and to Avoid Public Health Failures," *Am J Public Health*, 2010 February, 100(2): 240–246.

[49] Russell Hotten, "Volkswagen: The scandal explained," *BBC News,* December 10, 2015, http://www.bbc.com/news/business-34324772, accessed December 2015.

[50] Russell Hotten, "Volkswagen: The scandal explained," *BBC News,* December 10, 2015, http://www.bbc.com/news/business-34324772, accessed December 2015.

[51] Healthy People 2020, "Environmental Health," *HealthyPeople.gov website,* http://www.healthypeople.gov/2020/topics-objectives/topic/environmental-health, accessed December 2015.

unmanned aerial systems—more commonly known as drones. Drones were aircraft that did not require a pilot to be present, and could be controlled remotely by the user. What's more, consumer drone sales were increasing rapidly in the US—with some predicting a rise from 1.9 million in 2016 4.3 million by 2020.[52]

After complaints of drones accidentally hitting people and flying too close to airplanes, discussions about drone safety and appropriate use increased.[53] While some advocated for better safety features, such as improved crash-avoidance technology, others advocated for additional regulation.[54] In the US, federal regulation around drones had been sparse until late in 2015, prompting state and local governments to create their own laws promoting safety.[55]

However, in 2015, the Federal Aviation Administration (FAA) set up the Unmanned Aerial Systems Task Force to create federal safety and use standards for drones.[56] In the fact sheet it released in December 2015, the agency discussed how the series of different laws in different places could make drone usage more dangerous—a contention that many local governments disagreed with.[57] Daniel Garodnick, a City Council member in New York City stated, "New York City is different from the cornfields of Iowa. That should be obvious to everyone, but that isn't reflected in F.A.A. rules."[58]

Unintended Consequences

Many consumer products and services had unintended public health effects. Most products and services outside of the healthcare industry were primarily created to solve a consumer need unrelated to health—and considering potential health risks was not always the first business priority. In August, 2014, public leaders told

[52] Staff writer, "Drones club," *The Economist,* April 23, 2016, http://www.economist.com/news/science-and-technology/21697214-better-technology-and-tougher-enforcement-rules-needed-safe, accessed May 2016.

[53] CECILIA KANG, "Drone Shopping? F.A.A. Rules May Hover Over Holidays," *The New York Times,* November 23, 2015, http://www.nytimes.com/2015/11/24/technology/proposed-regulations-for-drones-are-released.html, accessed December 2015.

[54] Staff writer, "Drones club," *The Economist,* April 23, 2016, http://www.economist.com/news/science-and-technology/21697214-better-technology-and-tougher-enforcement-rules-needed-safe, accessed May 2016.

[55] CECILIA KANG, "F.A.A. Drone Laws Start to Clash With Stricter Local Rules," *The New York Times,* December 27, 2015, http://www.nytimes.com/2015/12/28/technology/faa-drone-laws-start-to-clash-with-stricter-local-rules.html?ref=business, accessed December 2015.

[56] Mark Harris, "Drone regulation is coming in time for Christmas, says FAA taskforce member," *The Guardian,* November 21, 2015, http://www.theguardian.com/technology/2015/nov/21/drone-regulation-before-christmas-says-faa-taskforce-member, accessed December 2015.

[57] CECILIA KANG, "F.A.A. Drone Laws Start to Clash With Stricter Local Rules," The New York Times, December 27, 2015, http://www.nytimes.com/2015/12/28/technology/faa-drone-laws-start-to-clash-with-stricter-local-rules.html?ref=business, accessed December 2015.

[58] Ibid.

residents of Toledo, Ohio to avoid using the tap water after tests revealed pollutants in the water supply.[59] Algae blooms, caused by warm temperatures, phosphorus, and nitrogen, were growing out of control in nearby Lake Erie and created toxins known to cause neurological deficits in humans.[60] The nitrogen and phosphorous predominantly came from agricultural fertilizers.[61] Though fertilizers can improve crop yield for farmers, their use had unintentionally left a city without safe drinking water.

Surfacing unintended consequences could be particularly challenging for newer technologies, whose risks were not always fully understood at first. Though headphones were first patented around 1890, modern headphones were not widely used until the late 1970s and early 1980s.[62] The Sony Walkman, a small, cassette tape player, was the first, affordable personal music player; it entered the US consumer market in 1979 and allowed people to listen to their music through headphones.[63] By 2015, one was hard pressed to find an American teenager that listened to their music via cassette tape; however, headphones remained commonplace and were used with smartphones and other personal audio players.

While consumers gained personal enjoyment from music on-the-go, headphones came with the risk of hearing loss, which could have negative consequences on physical and mental health, as well as education and employment.[64] In 2015, the World Health Organization (WHO) published a press release discussing how 1.1 billion teenagers and young adults were at risk of hearing loss due to the unsafe use of personal audio devices and exposure to damaging levels of sound at noisy entertainment venues (e.g., nightclubs, bars and sporting events).[65] Dr. Etienne Krug, WHO Director for the Department for Management of Noncommunicable Diseases, Disability, Violence and Injury Prevention stated,

> As they go about their daily lives doing what they enjoy, more and more young people are placing themselves at risk of hearing loss. They should be aware that once you lose your hearing, it won't come back. Taking simple preventive actions will allow people to continue to enjoy themselves without putting their hearing at risk.[66]

---

[59] Emma Fitzsimmons, "Tap Water Ban for Toledo Residents," *The New York Times,* August 3, 2014, http://www.nytimes.com/2014/08/04/us/toledo-faces-second-day-of-water-ban.html, accessed January 2016.

[60] Jane J. Lee, "Driven by Climate Change, Algae Blooms Behind Ohio Water Scare Are New Normal," *National Geographic,* August 6, 2014, http://news.nationalgeographic.com/news/2014/08/140804-harmful-algal-bloom-lake-erie-climate-change-science/, accessed January 2016.

[61] Ibid.

[62] Jimmy Stamp, "A Partial History of Headphones," *Smithsonian.com,* March 19, 2013, http://www.smithsonianmag.com/arts-culture/a-partial-history-of-headphones-4693742/?no-ist, accessed December 2015.

[63] Ibid.

[64] World Health Organization, "1.1 billion people at risk of hearing loss," *WHO Media centre,* February 27, 2015, http://www.who.int/mediacentre/news/releases/2015/ear-care/en/, accessed December 2015.

[65] Ibid.

[66] Ibid.

Lobbying

By 2015, businesses spent about $2.6 billion a year on lobbying activities in the US, causing some to cite concerns that corporate lobbying power could unduly impact policy-making activities to advance corporate interests.[67] A 2015 article in *The Atlantic* compared that $2.6 billion a year to what it costs to fund the House of Representatives and the Senate, $1.18 billion and $860 million a year respectively, showing that corporations spent more on lobbying than the country did on funding Congress.[68] The US was not the only country to contend with this issue in recent years. In 2010, David Cameron expressed similar concerns about corporate lobbying activities in the United Kingdom (UK), stating:

> We all know how it works. The lunches, the hospitality, the quiet word in your ear, the ex-ministers and ex-advisers for hire, helping big business find the right way to get its way...I believe that secret corporate lobbying, like the expenses scandal, goes to the heart of why people are so fed up with politics. It arouses people's worst fears and suspicions about how our political system works.[69]

Corporate lobbying provided numerous benefits to businesses—from improved market value to fewer regulatory challenges—and many corporations that were dependent on regulatory decisions had a lobbying strategy.[70] Corporate lobbying did not always lead to abuse, and lobbyists often conducted detailed and specific research for policy makers. One *Business Insider* article stated:

> The increasing complexity of policy also makes it more difficult for generalist and generally inexperienced government staffers to maintain an informed understanding of the rules and regulations they are in charge of writing and overseeing. They typically have neither the time to specialize nor the experience to draw on. As a result, staffers must rely more and more on the lobbyists who specialize in particular policy areas. This puts those who can afford to hire the most experienced and policy-literate lobbyists—generally large companies—at the center of the policymaking process.[71]

Despite the expertise benefits it provided, corporate lobbying could also negatively impact consumer health. One of the earliest examples of this was in the lead

---

[67] Lee Drutman, "How Corporate Lobbyists Conquered American Democracy," *The Atlantic,* April 20, 2015, http://www.theatlantic.com/business/archive/2015/04/how-corporate-lobbyists-conquered-american-democracy/390822/, accessed December 2015.

[68] Ibid.

[69] Andrew Sparrow, "David Cameron vows to tackle 'secret corporate lobbying'," *The Guardian,* February 8, 2010, http://www.theguardian.com/politics/2010/feb/08/david-cameron-secret-corporate-lobbying, accessed December 2015.

[70] Steven Strauss, "Here's Everything You've Always Wanted To Know About Lobbying For Your Business," *Business Insider,* November 25, 2011, http://www.businessinsider.com/everything-you-always-wanted-to-know-about-lobbying-2011-11, accessed December 2015.

[71] New America's Weekly Wonk: Lee Drutman, "How corporations turned into political beasts," *Business Insider,* April 25, 2015, http://www.businessinsider.com/how-corporations-turned-into-political-beasts-2015-4, accessed December 2015.

paint industry. Lead paint was commonly used prior to the 1970s.[72] It posed significant health risks to those who accidentally ingested it—particularly for children who might eat paint chips or get paint dust on their hands or toys. By the 2000s, it was widely accepted that lead paint damaged the brain, kidneys, nerves and blood; it could also cause behavioral problems, learning disabilities, seizures and death.[73] However, in the 1950s, when local public health officials in several cities began to realize the damaging effects lead paint could have, lobbyists succeeded in disarming their attempts to restrict its use or warn consumers of its risks.[74] Consumer use of lead paint was not banned by the federal government until 1978, about two decades after the first actions of local health officials.[75]

Similarly, US food manufacturers invariably lobbied against any new regulations, and even against new, voluntary standards.[76] In more recent history, many physicians, nutritionists, and public health professionals argued that the US government was overly influenced by the food industry when it developed and released the 2015 Dietary Guidelines for Americans.[77] The guidelines were important because they not only affected the everyday eating habits of Americans, but they also dictated a wide array of policies and programs (e.g., food label requirements, school lunch decisions, public nutrition programs, and research grants).[78]

The American Cancer Society Cancer Action Network, an arm of the American Cancer Society, released a written statement in response to the guidelines.[79] The statement applauded the guideline's efforts to reduce sugar consumption, but

---

[72] DAVID ROSNER AND GERALD MARKOWITZ, "Why It Took Decades of Blaming Parents Before We Banned Lead Paint," *The Atlantic,* April 22, 2013, http://www.theatlantic.com/health/archive/2013/04/why-it-took-decades-of-blaming-parents-before-we-banned-lead-paint/275169/, accessed December 2015.

[73] U.S. Department of Housing and Urban Development, "About Lead-Based Paint," *HUD.gov website,* http://portal.hud.gov/hudportal/HUD?src=/program_offices/healthy_homes/healthyhomes/lead, accessed December 2015.

[74] DAVID ROSNER AND GERALD MARKOWITZ, "Why It Took Decades of Blaming Parents Before We Banned Lead Paint," *The Atlantic,* April 22, 2013, http://www.theatlantic.com/health/archive/2013/04/why-it-took-decades-of-blaming-parents-before-we-banned-lead-paint/275169/, accessed December 2015.

[75] Ibid.

[76] Lyndsey Layton and Dan Eggen, "Industries lobby against voluntary nutrition guidelines for food marketed to kids," *The Washington Post,* July 9, 2011, https://www.washingtonpost.com/politics/industries-lobby-against-voluntary-nutrition-guidelines-for-food-marketed-to-kids/2011/07/08/gIQAZSZu5H_story.html, accessed May 2016.

[77] Markham Heid, "Experts Say Lobbying Skewed the U.S. Dietary Guidelines," *Time,* January 8, 2016, http://time.com/4130043/lobbying-politics-dietary-guidelines/, accessed February 2016.

[78] Ibid.

[79] American Cancer Society Cancer Action Network, "New Dietary Guidelines Disregard Important Link Between Diet and Cancer; Missed Opportunity to Reduce Death and Suffering," *American Cancer Society Cancer Action Network Media Center*, January 7, 2016, http://www.acscan.org/content/media-center/new-dietary-guidelines-disregard-important-link-between-diet-and-cancer-missed-opportunity-to-reduce-death-and-suffering/, accessed February 2016.

expressed frustration with its position on red meat.[80] It quoted Dr. Richard Wender, chief cancer control officer of the American Cancer Society, who stated, "The science on the link between cancer and diet is extensive. By omitting specific diet recommendations, such as eating less red and processed meat, these guidelines miss a critical and significant opportunity to reduce suffering and death from cancer…"[81]

What Is "Healthy"?

By 2015, it was clear that many consumers wanted "healthier" products—especially in the food industry—and some were willing to pay more for them.[82] However, regulators, corporations, and consumers themselves weren't always quite sure what that meant—health was complicated. Furthermore, consumer education was often required to change behavior, and it was not uncommon for consumers to find conflicting information about an industry or product. Consider this phenomenon in the food industry. Terms such as "local", "organic", "natural", and "non-GMO" (genetically modified organism) flooded the food market, making the health benefits of each confusing. Furthermore, many of them, including "natural" and "local" did not have federally mandated definitions.[83,84]

In response to consumer demand for more natural ingredients, many food companies began removing additives from their foods.[85] However, after Panera Bread announced that it would remove dyes and additives from its food, Michael Jacobson, the Executive Director of the Center for Science in the Public Interest, a consumer advocacy group, stated "But just because something is artificial or its name is hard to pronounce doesn't mean it's unsafe. Some of the additives Panera is ditching are perfectly innocuous, such as calcium propionate or sodium lactate—so those moves are more about public relations than public health."[86]

---

[80] Ibid.

[81] Ibid.

[82] Nancy Gagliardi, "Consumers Want Healthy Foods—And Will Pay More For Them," *Forbes,* February 18, 2015, http://www.forbes.com/sites/nancygagliardi/2015/02/18/consumers-want-healthy-foods-and-will-pay-more-for-them/, accessed December 2015.

[83] Food and Drug Administration (FDA), "What is the meaning of 'natural' on the label of food?," *FDA website,* accessed at http://www.fda.gov/aboutfda/transparency/basics/ucm214868.htm, accessed January 2016.

[84] United States Department of Agriculture Economic Research Service, "Local Foods," *USDA website,* accessed at http://www.ers.usda.gov/topics/food-markets-prices/local-foods.aspx, accessed January 2016.

[85] Christopher Doering, "Consumers demand healthier ingredients," *USA Today, April 3, 2015,* http://www.usatoday.com/story/money/business/2015/04/03/companies-respond-to-demand-for-healthier-ingredients/25250867/, accessed December 2015.

[86] Michael F. Jacobson, "Panera Removes Dyes, Additives from Foods," *Center for Science in the Public Interest website,* May 5, 2015, http://www.cspinet.org/new/201505051.html, accessed December 2015.

Researchers had found that "organic" food did not necessarily denote better nutritional quality either.[87] "You can't use organic as your sole criteria for judging nutritional quality," Crystal Smith-Spangler, a physician and researcher said.[88] Vegetables did indeed vary in their nutritional quality, but this was also affected by their genetic variety, the soil quality, the time at which they were picked, and the weather.[89]

By 2016, a large number of foods contained genetically modified ingredients. For example, many seeds were genetically modified for resistance to herbicides.[90] Though many consumers fought the use of these ingredients and demanded the right to know what was in their food, the U.S. Food and Drug Administration had deemed GMOs safe and the federal government did not require extra labeling for foods that contained such ingredients.[91]

While the federal government remained silent on required GMO-labels, in 2016, it appeared likely that some states would pass their own labelling laws. Vermont was poised to require GMO labels for many foods beginning on July 1, 2016. Though the law included exemptions—food in restaurants, as well as meats and cheeses, would not require labels—the law was likely to have far-reaching effects. One article stated:

> …[the law] will have nationwide implications. Because food manufacturers may not want to create separate packaging for different regions of the country, or to risk the legal liability if a non-labeled GMO winds up in Vermont, they will probably adjust their supply chains far beyond New England.[92]

As a result of such concerns, in 2016, Campbell Soup Company ("Campbell's"), which had previously opposed mandated GMO labelling on food packaging, announced its support for federally-mandated GMO labelling.[93] In a statement on its website, Campbell's President and CEO Denise Morrison reiterated that Campbell's

---

[87] Crystal Smith-Spangler, Margaret L. Brandeau, Grace E. Hunter, J. Clay Bavinger, Maren Pearson, Paul J. Eschbach, Vandana Sundaram, Hau Liu, Patricia Schirmer, Christopher Stave, Ingram Olkin, and Dena M. Bravata, "Are Organic Foods Safer or Healthier Than Conventional Alternatives?: A Systematic Review," *Ann Intern Med.,* 2012;157(5):348–366.

[88] Allison Aubrey and Dan Charles, "Why Organic Food May Not Be Healthier For You," *NPR,* September 4, 2012, http://www.npr.org/sections/thesalt/2012/09/04/160395259/why-organic-food-may-not-be-healthier-for-you, December 2015.

[89] Ibid.

[90] Mary Clare Jalonick, "Food industry pushing halt to labeling of genetically modified items," *The Boston Globe,* December 9, 2015, https://www.bostonglobe.com/business/2015/12/08/food-industry-pushing-halt-gmo-labeling-end-year/rtVa7JbI7oQg3UfkhiLINJ/story.html, accessed December 2015.

[91] Ibid.

[92] JAYSON LUSK, "Can I Get That With Extra GMO?," *The Wall Street Journal,* April 26, 2016, http://www.wsj.com/articles/can-i-get-that-with-extra-gmo-1461710531, accessed May 2016.

[93] Campbell Team, "WHY WE SUPPORT MANDATORY NATIONAL GMO LABELING," *Campbell's website,* January 7, 2016, accessed at http://www.campbellsoupcompany.com/newsroom/news/2016/01/07/labeling/, accessed January 2016.

does not dispute the science behind GMOs and believes in their safety; however, she explained that some states had passed their own regulations around GMO labelling on food packaging, leading to a patchwork of state laws across the country.[94] She stated, "Put simply, although we believe that consumers have the right to know what's in their food, we also believe that a state-by-state piecemeal approach is incomplete, impractical and costly to implement for food makers. More importantly, it's confusing to consumers."[95]

Standards differed in the European Union, where stricter labeling was required for foods that contained GMOs.[96] The Food Standards Agency, a government department in the UK, stated on its website: "The Agency supports consumer choice. We recognise that some people will want to choose not to buy or eat genetically modified (GM) foods, however carefully they have been assessed for safety."[97]

How Corporations Pursue Consumer Health

Corporations deliberately pursued consumer health in different ways depending on their product or service, consumers, and their corporate competencies and vulnerabilities. Corporate actions to advance consumer health broke down into five main categories. Some corporations took actions across several areas, while others focused efforts in one category. These categories were:

1. **Enhancing product or service quality**: Companies enhanced product and service quality through innovative design strategies, healthier ingredients and raw materials, improved supply chain management, manufacturing and safety requirements, and redesigned packaging.
2. **Re-evaluating sales and distribution decisions**: Businesses used sales and distribution strategies—such as pricing schemes, directed advertising, and narrow distribution channels—to ensure that the right consumers used products for their intended purposes.
3. **Improving consumer use**: Businesses improved customers' use of their products through detailed instruction, ongoing customer support, and guidance for appropriate use.
4. **Advocating for improved disposal**: Companies advocated for more sustainable and healthful means of product disposal.
5. **Advancing broader industry standards**: Corporations advanced industry standards to promote consumer health.

[94] Ibid.

[95] Ibid.

[96] Food Standards Agency (FSA), "GM Labelling," *FSA website* https://www.food.gov.uk/science/novel/gm/gm-labelling, accessed December 2015.

[97] Ibid.

Enhancing Product or Service Quality

Design

One of the most common ways that corporations enhanced the health and safety of their products and services was through product and service design. The Lego Group, a children's toy manufacturer based in Denmark, had $4.7B in total revenue in 2014, making it one of the leading toy manufacturers in the world.[98] While Lego produced both physical and digital products, all products were based on the "the physical brick experience," usually requiring children to build the toys themselves from smaller pieces (often bricks).[99]

Its physical toy products, the core of its business, had the potential to be health-promoting and health-harming for children. The toys increased cognitive development and dexterity in young children, which could be considered especially important as many cited concerns that more and more children lacked fine motor skills due to the increased use of tablet computers, smartphones, and other technology for entertainment.[100] However, because many of the toys had a large number of pieces, there were some concerns that children, especially young ones, could ingest the pieces. Despite these concerns, in 2014, Lego had not had any product recalls for the five years prior.[101]

Its success could be largely attributed to a company-wide focus on safety, which included robust product planning and development. (See Exhibit 2.3 for Lego's Safety Assessment.) On its website, Lego discussed how it began its pursuit of safety early in the design process:

> We see product safety as the responsibility not only of product safety specialists, but of all involved in the development process…Product safety specialists interacted with product designers at the earliest opportunity to ensure that every design is thoroughly evaluated and assessed in terms of product safety. Product designers are continuously trained in the principles of product safety to understand how to proactively incorporate safety into their designs.[102]

---

[98] Richard Milne, "Lego Enters a New Dimension with its Digital Strategy," *Financial Times,* September 27, 2015, http://www.ft.com/intl/cms/s/0/7fa681e2-635d-11e5-a28b-50226830d644.html#axzz3tqZUmxmy, accessed December 2015.

[99] Ibid.

[100] Graeme Paton, "Infants 'unable to use toy building blocks' due to iPad addiction," *The Telegraph,* April 15, 2014, http://www.telegraph.co.uk/education/educationnews/10767878/Infants-unable-to-use-toy-building-blocks-due-to-iPad-addiction.html, accessed December 2015.

[101] Lego, "SAFE, HIGH-QUALITY PRODUCTS," *Lego website,* http://www.lego.com/en-us/aboutus/responsibility/responsibilityreport2014/impactofthebrick/safe-high-quality-products, accessed December 2015.

[102] Ibid.

Exhibit 2.3: The LEGO Group Safety Assessment

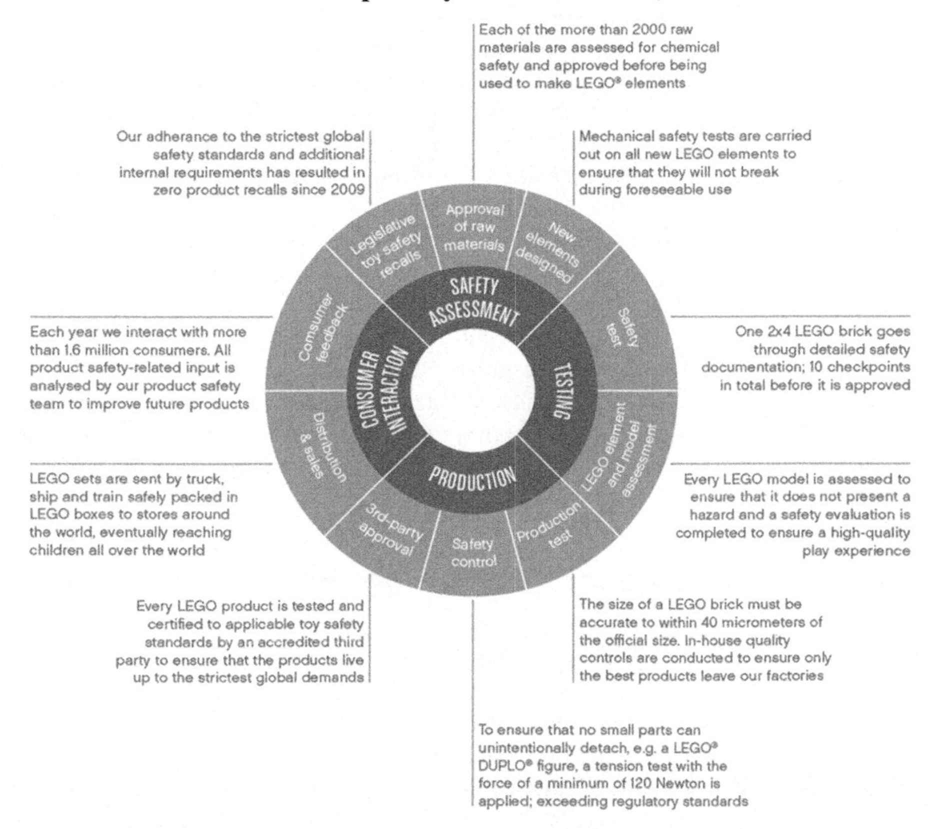

Source: Lego, "SAFE, HIGH-QUALITY PRODUCTS," *Lego website,* http://www.lego.com/en-us/aboutus/responsibility/responsibilityreport2014/impactofthebrick/safe-high-quality-products, accessed December 2015.

Improving design to improve safety applied not only to consumer products, but also to services. One service area where this concept had become increasingly important was in hospitals. Though people typically went to hospitals to improve their health when they were ill, hospital stays carried the potential to become health-harming (e.g., hospital-acquired infections, patient falls, and psychological distress).[103] Patient falls were of particular concern for hospitals, as researchers had found that a patient fall could increase hospital costs by around $13,000 and length of stay by over 6 days.[104] Research showed that about 30 % of patients who fell were

[103] Robert Pearl, "4 Ways Hospitals Can Harm You," *Forbes,* January 24, 2014, http://www.forbes.com/sites/robertpearl/2014/01/24/4-ways-hospitals-can-harm-you/, accessed December 2015.

[104] CA Wong, AJ Recktenwald, ML Jones, BM Waterman, ML Bollini, WC Dunagan, "The cost of serious fall-related injuries at three Midwestern hospitals," *Jt Comm J Qual Patient Saf.*, 2011 Feb;37(2):81–7.

injured and since 2008, the Centers for Medicare and Medicaid Services (CMS) had not reimbursed providers for hospital stays that included a patient fall.[105]

In response, hospitals redesigned processes and even altered the physical environment to reduce fall risks. In 2012, the University Medical Center of Princeton at Plainsboro opened an entirely new hospital in New Jersey. The medical center had planned the new hospital for several years prior, and as part of these efforts, it had created "mock up" patient rooms at its old location to test the effects of different room designs. When the medical center opened at the new location in 2012, it incorporated the most effective design changes to promote safety, including several specifically aimed at reducing patient falls (e.g., handrails from patients' beds to their bathrooms, bathroom doors that opened easily, patient beds that could be lowered to 16 in. off the floor, and a system for contacting nurses if high-risk patients got up without help).[106] Barry Rabner, the president and CEO of Princeton HealthCare System discussed the positive results, stating, "We had clear data on our performance in the old building and have been rigorous in tracking it on an ongoing basis in the new building. And all of the measures have improved materially."[107]

Materials and Ingredients

Many corporations improved the health and safety of their products by using health-promoting materials or ingredients. This trend was evident in the food and beverage industry, where throughout the late 2000s consumers demanded greater transparency and more natural ingredients.[108] Market research in 2014 had shown that more than 60 % of Americans stated that the absence of artificial colors or flavors was important to their food purchase decisions.[109] Many in the industry called this trend a push for a "cleaner label."[110]

In 2015, General Mills, a US manufacturer and marketer of branded consumer foods, announced that it planned to phase out artificial flavors from its cereals by 2017.[111] At the time, General Mills had 20 % of its business in cereal, and was well-

[105] Jennifer Goldsack, Janet Cunningham, Susan Mascioli, "Patient falls: Searching for the elusive 'silver bullet'," *Nursing, July 2014,* Volume 44, Issue 7, pp. 61–62.

[106] "How to Design Hospitals With Safety In Mind," *Hospitals &Health Networks Magazine,* October 14, 2014, http://www.hhnmag.com/articles/3934-how-to-design-hospitals-with-safety-in-mind, accessed December 2015.

[107] Ibid.

[108] Christopher Doering, "Consumers demand healthier ingredients," *USA Today, April 3, 2015,* http://www.usatoday.com/story/money/business/2015/04/03/companies-respond-to-demand-for-healthier-ingredients/25250867/, accessed December 2015.

[109] Ibid.

[110] Ibid.

[111] James Hamblin, "Lucky Charms, the New Superfood," *The Atlantic,* June 23, 2015, http://www.theatlantic.com/health/archive/2015/06/general-mills-to-phase-out-artificial-cereal-dyes/396536/, accessed December 2015.

known for brands such as Lucky Charms and Trix, which had used artificial flavors and ingredients,[112,113] General Mills president, Jim Murphy, said, "People don't want colors with numbers in their food anymore."[114] The move was not the first by General Mills to appease health-conscious consumers. In 2014, it had announced that its "Original Cheerios" were GMO-free.[115] A large number of other food manufacturers and restaurants, including Subway, Pizza Hut, Panera, and Hershey made similar announcements and vowed to reduce artificial ingredients in their products.[116]

Artificial ingredients were not the only targets—many corporations also sought to reduce fat and sugar in their products. In 2013, The Dannon Company ("Dannon") made a 3-year commitment in partnership with The Partnership for a Healthier America (PHA), a nonpartisan nonprofit organization that works with the private sector on mitigating the childhood obesity crisis. Dannon pledged to not only improve the nutrient density of its product portfolio (e.g., increase amount of Vitamin D), but also to reduce the amount of fat and sugar in its products by June, 2016.[117] In 2015, Dannon announced that it had reached its goal to reduce the amount of total sugar in its products to 23 g or less per 6-oz serving in all products for children and 70 % of its overall products.[118]

Some companies went a step further to not only remove harmful ingredients, but to reengineer the core purpose of their products. Nestlé, a food and drink company known for confectionary and frozen dinners, was on a mission to redefine itself as a scientifically driven wellness company.[119] In 2016, a *Bloomberg Businessweek* article showcased how the company hoped to soon offer medical treatments—delivered through food.[120] By 2016, the Nestlé Institute of Health Sciences employed over 160 scientists focused on new product development. Nestlé Health Science, a sub-

---

[112] General Mills, "Company Overview: Making food with passion for 150 years," *General Mills website,* https://www.generalmills.com/en/Company/Overview, accessed December 2015.

[113] James Hamblin, "Lucky Charms, the New Superfood," *The Atlantic,* June 23, 2015, http://www.theatlantic.com/health/archive/2015/06/general-mills-to-phase-out-artificial-cereal-dyes/396536/, accessed December 2015.

[114] Ibid.

[115] Tom Forsythe, "The One and Only Cheerios," *taste of General Mills blog,* accessed at http://www.blog.generalmills.com/2014/01/the-one-and-only-cheerios/, accessed January 2016.

[116] The Associated Press, "No More Artificial Colors for Trix or Reese's Puffs," *The New York Times,* June 22, 2015, http://www.nytimes.com/2015/06/23/business/no-more-artificial-colors-for-trix-or-reeses-puffs.html, accessed December 2015.

[117] Partnership For A Healthier America, "Dannon," *Partnership For A Healthier America website,* http://ahealthieramerica.org/our-partners/#Dannon,5578,partner hide, accessed December 2015.

[118] PR Newswire, "The Dannon Company Makes Significant Year One Progress Toward Partnership For A Healthier America Commitment," February 25, 2015, http://www.prnewswire.com/news-releases/the-dannon-company-makes-significant-year-one-progress-toward-partnership-for-a-healthier-america-commitment-300041131.html, accessed December 2015.

[119] Matthew Campbell and Corinne Gretler, "Nestlé Wants to Sell You Both Sugary Snacks and Diabetes Pills," *Bloomberg Businessweek,* May 5, 2016, http://www.bloomberg.com/news/features/2016-05-05/nestl-s-sugar-empire-is-on-a-health-kick, accessed May 2016.

[120] Ibid.

sidiary with more than 3000 employees, commercialized these new, healthier products.[121] Some would likely require a prescription, while others would be sold over-the-counter.[122] Though questions remained regarding the regulation of such products, *Bloomberg Businessweek* wrote:

> If this vision is realized, a visit to the family doctor in a decade's time might end with a prescription for a tasty Nestlé shake for heart trouble or a recommendation for an FDA-approved tea to strengthen aging joints. The company would expand from the vending machine and supermarket to the pharmacy, doctor's office, and hospital. At the same time, it would keep its core food and sweets businesses. In other words, Nestlé would sell a problem with one hand and a remedy with the other.[123]

Supply Chain Management

In the food industry, many companies increased vigilance around supply chain management due to consumer and regulator concerns around contaminants in food—especially as food industries matured in emerging markets and the food supply chain globalized.[124] In 2015, Dr. Margaret Chan, Director-General of the World Health Organization (WHO), stated:

> Food production has been industrialized and its trade and distribution have been globalized. These changes introduce multiple new opportunities for food to become contaminated with harmful bacteria, viruses, parasites, or chemicals. A local food safety problem can rapidly become an international emergency. Investigation of an outbreak of foodborne disease is vastly more complicated when a single plate or package of food contains ingredients from multiple countries.[125]

Consumer concerns around food safety were not new; however, they had increased throughout the late 2000s due to several high-profile food recalls and a new consumer focus on understanding ingredients and potential allergens (e.g., gluten, dairy) in foods.[126] In 2007, regulators recalled pet food when 14 pets died from food contaminated with melamine that had been manufactured in China.[127] Melamine also contaminated milk products in China in 2008, killing six infants and making more than 300,000 ill.[128] In addition, many consumers sought confirmation

[121] Ibid.

[122] Ibid.

[123] Ibid.

[124] John A. Quelch and Margaret L. Rodriguez, "Mérieux NutriSciences: Marketing Food Safety," HBS No. 516-024 (Boston: Harvard Business School Publishing, 2015).

[125] World Health Organization, "World Health Day 2015: From farm to plate, make food safe," *WHO website: Media centre,* April 2, 2015, http://www.who.int/mediacentre/news/releases/2015/food-safety/en/, accessed December 2015.

[126] John A. Quelch and Margaret L. Rodriguez, "Mérieux NutriSciences: Marketing Food Safety," HBS No. 516-024 (Boston: Harvard Business School Publishing, 2015).

[127] Ibid.

[128] Ibid.

of "authenticity" from their food—that is, what they were eating was what they thought they were eating. In 2013, the media reported that frozen burgers from UK retailer, Tesco, labelled as beef contained horse meat.[129]

However, it wasn't only global supply chains that presented challenges; sometimes, using more local ingredients also created issues. Chipotle Mexican Grill Inc. ("Chipotle"), a fast-casual Mexican restaurant, prided itself on fresh ingredients and producing "food with integrity."[130] Unlike many other fast-food chains, it used local ingredients and fresh produce.[131] However, in 2015, several disease outbreaks showcased that scaling the local food movement could be difficult and even detrimental to public health. It experienced an *E. coli* outbreak in nine states, sickening over 50 people in total.[132] In a separate incident, more than 140 Boston College students caught the norovirus after eating at a Chipotle near their campus.[133] The chain was forced to reevaluate its use of local ingredients—especially fresh produce. Following these incidents, media reports stated that Chipotle was expected to lower its use of locally-sourced ingredients and centralize the preparation of most vegetables.[134] Many were quick to point out that Chipotle was moving its practices in the same direction as many of the other fast-food companies that it had previously criticized.[135]

Safety Testing

Corporations across many kinds of industries pursued consumer health through enhanced manufacturing processes and product testing. For many of the same reasons that they invested in supply chain management, food and beverage companies also pursued enhanced safety testing. Regulators in the US supported growth in food testing and quality assurance. The FDA stated that it passed the Food Safety

---

[129] Felicity Lawrence, "Horsemeat scandal: where did the 29% horse in your Tesco burger come from?," *The Guardian,* October 22, 2013, http://www.theguardian.com/uk-news/2013/oct/22/horsemeat-scandal-guardian-investigation-public-secrecy, accessed December 2015.

[130] Chipotle, "Food with Integrity," *Chipotle website,* https://www.chipotle.com/food-with-integrity, accessed December 2015.

[131] Julie Jargon, "Chipotle Pulls Back on Local Ingredients," *The Wall Street Journal,* December 15, 2015, http://www.wsj.com/articles/chipotle-heads-back-to-the-test-kitchen-1450205438, accessed December 2015.

[132] Ibid.

[133] Kristi Palma, "Everything you need to know about norovirus, the illness afflicting Boston College students," *Boston.com,* December 10, 2015, http://www.boston.com/health/2015/12/10/everything-you-need-know-about-norovirus-the-illness-afflicting-boston-college-students/u2SdZNUKSk6rgZKyqA53VI/story.html, accessed December 2015.

[134] Julie Jargon, "Chipotle Pulls Back on Local Ingredients," *The Wall Street Journal,* December 15, 2015, http://www.wsj.com/articles/chipotle-heads-back-to-the-test-kitchen-1450205438, accessed December 2015.

[135] Ibid.

Modernization Act (FSMA)—the most extensive reform in food safety laws in 70 years—in 2011 to "ensure the US food supply [was] safe by shifting the focus from responding to contamination to preventing it."[136]

Because of this, the global food testing market was poised to grow; it was roughly $3.5 billion in 2014 and was projected to reach $4.6 billion by 2018.[137] Companies that invested in food testing included food manufacturers, retailers (e.g., grocery and drug stores), and restaurants; while some of these corporations kept such testing in-house, increasingly, many others opted to outsource these responsibilities to food safety and testing companies.[138]

While implementing safety testing was a crucial first step in promoting consumer safety, safety testing on its own was not enough—businesses also needed adequate reporting protocols for safety issues identified during the testing process. In February 2014, General Motors ("GM") issued recalls for over 2 million cars that had faulty ignition switches, causing the car's engine to shut off while still in motion and disabling air bags from deploying.[139] Interestingly, GM admitted that several employees knew about the problem—it was first identified in 2001.[140] In 2004, GM engineers had suggested a solution to the faulty switch, but executives decided against implementing it due to concerns around the lead time required, cost and effectiveness.[141]

This inadequate level of reporting had disastrous consequences. Reports estimated that the defect caused over 120 deaths and 275 injuries.[142] In 2014, the National Highway Traffic Safety Administration (NHTSA) required GM to pay a $35 million civil penalty for its delayed reporting of the defect.[143] In 2015, GM paid $900 million to settle federal charges and another $575 million to settle civil lawsuits.[144] GM's CEO Mary Barra said, "I have said many times I wish I could turn

---

[136] U.S. Food and Drug Administration, "FDA Food Safety Modernization Act (FSMA)," *FDA website,* http://www.fda.gov/Food/GuidanceRegulation/FSMA/, accessed December 2015.

[137] John A. Quelch and Margaret L. Rodriguez, "Mérieux NutriSciences: Marketing Food Safety," HBS No. 516-024 (Boston: Harvard Business School Publishing, 2015).

[138] Ibid.

[139] Tanya Basu, "Timeline: A History Of GM's Ignition Switch Defect," *NPR,* March 31, 2014, http://www.npr.org/2014/03/31/297158876/timeline-a-history-of-gms-ignition-switch-defect, accessed December 2015.

[140] Ibid.

[141] Christopher Jensen, "In General Motors Recalls, Inaction and Trail of Fatal Crashes," *The New York Times,* March 2, 2014, http://www.nytimes.com/2014/03/03/business/in-general-motors-recalls-inaction-and-trail-of-fatal-crashes.html, accessed December 2015.

[142] Jeff Glor, "GM CEO: 'I wish I could turn back the clock'," *CBS News,* September 17, 2015, http://www.cbsnews.com/news/gm-ceo-i-wish-i-could-turn-back-the-clock/, accessed December 2015.

[143] Tanya Basu, "Timeline: A History Of GM's Ignition Switch Defect," *NPR,* March 31, 2014, http://www.npr.org/2014/03/31/297158876/timeline-a-history-of-gms-ignition-switch-defect, accessed December 2015.

[144] Jeff Glor, "GM CEO: 'I wish I could turn back the clock'," *CBS News,* September 17, 2015, http://www.cbsnews.com/news/gm-ceo-i-wish-i-could-turn-back-the-clock/, accessed December 2015.

back the clock. If we could we would but the only thing we can do is make sure we respond in the right way and we have done that in the case and we continue to do so in everything."[145]

Packaging

Packaging—the means by which a product was stored, shipped, and sold– could change its health effects on consumers. While some packaging decisions were purely made for aesthetic reasons, packaging was also sometimes used to control serving size or dose, promote safety (e.g., tamper-proof packaging), or preserve freshness (e.g., controlled atmosphere packaging).

Many food and beverage companies changed their packaging to control serving sizes. Some in the beverage industry created smaller bottles or cans to appeal to consumers looking for healthier options. Regan Ebert, the senior vice president for marketing at Dr. Pepper Snapple Group said, "There's consumers out there that don't want to consume too much. [The small cans] do a good job of solving that need for consumers."[146]

Other industries also utilized innovations in packaging to promote consumer health. In 2012, single-load laundry detergent pods hit the market.[147] The pods were often small and filled with brightly colored detergent, with a thin outer layer that dissolved when wet. However, this appearance closely resembled candy, causing many children, and some seniors with dementia, to ingest the pods.[148] (See Exhibit 2.4 for a picture of a laundry pod.) By mid-2015, more than 33,000 children aged five and under in the US had been exposed to the contents of laundry packets since 2012.[149] Though some children became so ill after ingestion that they required intubation and more intensive care, the most common reactions were vomiting, lethargy, irritated eyes, and coughing or choking.[150]

---

[145] Ibid.

[146] Margot Sanger-Katz, "The Decline of Big Soda," *The New York Times,* October, 4, 2015, http://www.nytimes.com/2015/10/04/upshot/soda-industry-struggles-as-consumer-tastes-change.html, accessed December 2015.

[147] Catherine Saint Louis, "Detergent Pods Pose Risk to Children, Study Finds," *The New York Times,* November 10, 2014, http://www.nytimes.com/2014/11/10/health/detergent-pods-pose-risk-to-children-study-finds.html, accessed December 2015.

[148] Serena Ng, "Detergent Makers to Add Bitter Substance to Laundry Pods," *The Wall Street Journal,* June 30, 2015, http://www.wsj.com/articles/laundry-pod-makers-take-steps-to-prevent-accidental-ingestion-1435692400, accessed December 2015.

[149] Ibid.

[150] Catherine Saint Louis, "Detergent Pods Pose Risk to Children, Study Finds," *The New York Times,* November 10, 2014, http://www.nytimes.com/2014/11/10/health/detergent-pods-pose-risk-to-children-study-finds.html, accessed December 2015.

Exhibit 2.4: Laundry Pod

Source: Casewriter.

In response, manufacturers made more secure closures and opaque packaging so that the contents were not visible; however, despite these efforts the accidental poisonings continued.[151] In 2015, European regulators required laundry pod manufacturers to use an aversive agent on the outside of the pods to make them taste bitter.[152] Procter & Gamble Co. ("P&G"), which not only operated in European markets, but also sold over 75 % of the laundry pods bought in the US, announced that it would also add a bitter-tasting substance to laundry pods sold in the US.[153] P&G also redesigned the packets to withstand the squeezing pressure of children, as well as delay release of the liquid detergent if ingested.[154] Although US regulators did not require the measures, as regulators in Europe did, the U.S. Consumer Product Safety Commission provided input into the standards and reported that it was monitoring the new safety measures for effectiveness.[155]

---

[151] Serena Ng, "Detergent Makers to Add Bitter Substance to Laundry Pods," *The Wall Street Journal,* June 30, 2015, http://www.wsj.com/articles/laundry-pod-makers-take-steps-to-prevent-accidental-ingestion-1435692400, accessed December 2015.

[152] Ibid.

[153] Ibid.

[154] Ibid.

[155] Ibid.

Re-evaluating Sales and Distribution Decisions

Corporations also used pricing, advertising, and distribution strategies to promote public health. Some corporations considered pricing strategies that included selling health-promoting products for lower prices in developing markets. Others used narrow or controlled distribution channels to ensure consumers accessed products appropriately. However, these kinds of decisions could also detract from consumer health. For example, some large food companies used pricing and distribution decisions to market unhealthy foods to young consumers, like children and teens.[156]

Colgate-Palmolive Company ("Colgate"), a leading provider of oral care products, considered such strategies in 2013 when it launched a new anti-cavity toothpaste, the Colgate® Maximum Cavity Protection™ plus Sugar Acid Neutralizer™.[157] In both developing and industrialized countries, cavities had remained a significant public health threat, and the new toothpaste formulation was clinically proven to reduce and prevent cavities more effectively than toothpaste with only fluoride.[158] Initially, Colgate created the product for sale in emerging markets, where it wanted to ensure that consumers could access it.[159] As the product launched, Colgate sought a balance between an affordable launch price to ensure an improved public health benefit and enough of a premium price over existing options to drive both credibility and profitability.[160]

Luvo, a food company that produced healthy frozen meals, improved marketing efficiency by leveraging its social media presence; then diverted the saved marketing dollars into improved product quality. To aid brand promotion, the company recruited several well-known brand ambassadors, including an Olympic swimmer and NFL quarterback to promote its products and serve as investors.[161] Ultimately, the company aimed to produce meals that were not only safe, minimally processed, low in sodium and sugar, with 1–3 servings of vegetables, but also affordable.[162] Despite a declining frozen food industry, the company had grown quickly and its products were sold in over 5000 grocery stores as of September, 2015.[163]

---

[156] KJ DELL'ANTONIA, More Research Suggests Fast-Food Advertising Works on Children," *The New York Times,* October 30, 2015, http://parenting.blogs.nytimes.com/2015/10/30/more-research-suggests-fast-food-advertising-works-on-children/, accessed May 2016.

[157] John A. Quelch and Margaret L. Rodriguez, "Colgate-Palmolive Company: Marketing Anti-Cavity Toothpaste," HBS No. 515-050 (Boston: Harvard Business School Publishing, 2015).

[158] Ibid.

[159] Ibid.

[160] Ibid.

[161] Staff Writer, "How Luvo Is Bringing The Heat To Frozen Foods," *Fast Company,* September 14, 2015, http://www.fastcompany.com/3050162/most-innovative-companies/how-luvo-is-bringing-the-heat-to-frozen-foods, accessed May 2016.

[162] Luvo, "Our Story: Nutrition," *Luvo Website,* http://luvoinc.com/our-story/nutrition/#E6Qyyb3hQ0EYBYgX.97, accessed May 2016.

[163] Staff Writer, "How Luvo Is Bringing The Heat To Frozen Foods," *Fast Company,* September 14, 2015, http://www.fastcompany.com/3050162/most-innovative-companies/how-luvo-is-bringing-the-heat-to-frozen-foods, accessed May 2016.

Many businesses took precautions to ensure that the right consumers accessed their products, making advertising and distribution decisions that promoted consumer health. A product could be safe or health-promoting for some individuals and health-harming for others. For example, in the medical marijuana industry, the drug could be a great benefit to those individuals with chronic pain; however, adolescent use of marijuana had been shown to cause several harmful long term outcomes.[164] Therefore, proper distribution channels were a necessity. Research in 2015 had shown that adolescent use of marijuana did not increase significantly after a state passed a medical marijuana law, meaning that just because a state had laws legalizing the medical usage of marijuana, teens were not using it more frequently.[165]

Improving Consumer Use

Many corporations sought to improve on how consumers used their products and services. To improve consumer use, corporations provided services such as detailed instructions or ongoing customer support. In 2016, Mars, a chocolate manufacturer, went a step further and began guiding the use of its products—advising consumers to eat some of its less healthy food products only once a week.[166] This strategy differed from that of other food manufacturers that opted to switch the ingredients in their products to healthier options. Because food companies could face significant backlash from unhappy customers when they changed their recipes—especially if they weren't transparent about it—encouraging reduced consumption with the same recipes offered an alternative strategy.[167] While some commended the company's efforts, others argued that food manufacturers had a responsibility to move consumer taste preferences towards healthier options.[168]

David Katz, the founding director of the Yale-Griffin Prevention Research Center, stated, "Taste buds prefer what they are used to. Food companies have played a central role in getting American taste buds used to …junk… but the process can be reverse engineered."[169]

[164] Deborah S Hasin, Prof Melanie Wall, Katherine M Keyes, Magdalena Cerdá, Prof John Schulenberg, Prof Patrick M O'Malley, Prof Sandro Galea, Prof Rosalie Pacula, Tianshu Feng, "Medical marijuana laws and adolescent marijuana use in the USA from 1991 to 2014: results from annual, repeated cross-sectional surveys," *The Lancet, July 2015, Volume 2, No. 7,* pp. 601–608.

[165] Ibid.

[166] Scheherazade Daneshkhu, "Big Food in Health Drive to keep Market Share," *Financial Times,* April, 26, 2016, http://www.ft.com/cms/s/0/83f05ea8-08a3-11e6-a623-b84d06a39ec2.html, accessed May 2016.

[167] Ibid.

[168] Ibid.

[169] Ibid.

23andMe, a personal genomics company that offered direct-to-consumer genetic testing, also faced serious challenges in the area of appropriate consumer use.[170] In order to achieve FDA-approval for its genetic tests and wellness reports, the company had to certify that it could convey this complicated, genetic information at a very basic reading level to consumers.[171] Because consumers might make health decisions based on the information, creating a foundational consumer understanding of genetics had become a critical part of the company's mission.

Advocating for Improved Disposal

Other businesses sought more healthful ways for consumers to dispose of products. As environmental conditions affected health, ensuring that consumers were able to recycle products contributed to consumer health.[172] Therefore, some corporations began to consider not only how consumers could appropriately use their products, but also how they could properly dispose of them.

Hewlett-Packard Company ("HP"), which provided computer hardware and software solutions, had developed extensive recycling and remanufacturing programs to reduce the environmental impact of product disposal. In 2015, it offered take-back programs in 73 countries and territories.[173] HP stated, "We are committed to helping our customers recycle responsibly, recovering 2.8 billion pounds of products since 1987." It also used "closed loop" recycled plastic in more than 75 % of its ink cartridges, meaning that the ink cartridges contained 50–70 % recycled plastic. HP also had remanufacturing programs, which aimed to refurbish hardware, thus reducing the environmental impact of disposal.[174]

While individual corporations created programs to promote recycling and appropriate disposal, recycling rates also varied across the world due to local recycling laws. For example, in the early 1990s, Germany passed an expansive recycling law.[175] A 1992 *Harvard Business Review* article on the topic stated:

> Germany's packaging ordinance, passed in April 1991, is probably the most ambitious environmental legislation any nation has ever attempted. It obliges retailers to take back

[170] 23andMe, "Reports," *23andMe website,* https://www.23andme.com/service/, accessed May 2016.

[171] Ibid.

[172] Healthy People 2020, "Environmental Health," *HealthyPeople.gov website,* http://www.healthypeople.gov/2020/topics-objectives/topic/environmental-health, accessed December 2015.

[173] HP, "Reuse & recycling at HP," *HP website: Sustainability,"* http://www8.hp.com/us/en/hp-information/environment/recycling-reuse.html, accessed December 2015.

[174] HP, "Reuse & recycling at HP," *HP website: Sustainability,"* http://www8.hp.com/us/en/hp-information/environment/recycling-reuse.html, accessed December 2015.

[175] Eliot Brown, "Germans Have a Burning Need for More Garbage," *The Wall Street Journal,* October 19, 2015, accessed at http://www.wsj.com/articles/germans-have-a-burning-need-for-more-garbage-1445306936, accessed January 2016.

> packaging from customers, manufacturers to retrieve it from retailers, and packaging companies to reclaim it from manufacturers.[176]

By 2015, Germany had one of the highest recycling rate worldwide—recycling around 65 % of household trash.[177] In 2013, about 34 % of waste in the US was recycled.[178]

Advancing Industry Standards

Many corporations advanced standards in their respective industries to promote consumer health. When a corporation shaped regulation to create a higher safety or health standard, it not only shaped its own destiny, but also often forced other companies in its industry to consider similar changes. In February 2014, CVS Caremark ("CVS") announced that it would stop selling tobacco products in its stores by October, 2014. To explain the decision, Larry Merlo, President and CEO of CVS, wrote:

> By removing tobacco products from our retail shelves, we will better serve our patients, clients and health care providers while positioning CVS Caremark for future growth as a health care company. Cigarettes and tobacco products have no place in a setting where health care is delivered. This is the right thing to do.[179]

The move to take tobacco off its shelfs was not only a boon for public health, it was also strategic for CVS. As part of the initiative, CVS ran an anti-smoking campaign, thus enabling the company to project a positive image to consumers and market anti-smoking aids.[180,181] The decision also allowed CVS to create better partnerships with local health systems and refocused its efforts on its core business—prescriptions.[182] Also in 2014, CVS announced that it would be changing its name

---

[176] Frances Cairncross, "How Europe's Companies Reposition to Recycle," *Harvard Business Review,* March-April 1992, accessed at https://hbr.org/1992/03/how-europes-companies-reposition-to-recycle, accessed January 2016.

[177] Eliot Brown, "Germans Have a Burning Need for More Garbage," *The Wall Street Journal,* October 19, 2015, accessed at http://www.wsj.com/articles/germans-have-a-burning-need-for-more-garbage-1445306936, accessed January 2016.

[178] United States Environmental Protection Agency, "Advancing Sustainable Materials Management," June 2015, accessed at http://www3.epa.gov/epawaste/nonhaz/municipal/pubs/2013_advncng_smm_fs.pdf, accessed January 2016.

[179] CVS Health, "Message from Larry Merlo, President and CEO," *CVS Health website,* February 5, 2014, https://www.cvshealth.com/newsroom/message-larry-merlo, accessed December 2015.

[180] Kyle Stock, "The Strategy Behind CVS's No-Smoking Campaign," *Bloomberg Business,* February 5, 2014, http://www.bloomberg.com/bw/articles/2014-02-05/the-strategy-behind-cvss-no-smoking-campaign, accessed January 2016.

[181] CVS Health, "Stop Smoking," *CVS website,* http://www.cvs.com/shop/health-medicine/stop-smoking/N-3tZ13ljjqZ2k, accessed January 2016.

[182] Kyle Stock, "The Strategy Behind CVS's No-Smoking Campaign," *Bloomberg Business,* February 5, 2014, http://www.bloomberg.com/bw/articles/2014-02-05/the-strategy-behind-cvss-no-smoking-campaign, accessed January 2016.

from "CVS Caremark" to "CVS Health" to reflect a broader commitment to its role in healthcare provision and innovation.[183] An article in *The New York Times* stated, "Its stand against smoking has allowed CVS to make alliances with health care providers and rebrand itself fully as a health care company. But with smoking rates on a steady decline, and cigarettes sales slumping, CVS also saw that future profits lie not with Big Tobacco but in health and wellness."[184]

This suggested that, resulting from better partnerships with health systems and increased prescription sales, total profit implications of the decision might be positive in the long-term. In February 2015, CVS positively announced that net revenues in the fourth quarter of 2014 increased 12.9 % to a record $37.1 billion.[185] However, in August 2015, CVS announced that while prescription sales continued to rise and second quarter year-over-year net revenue increased 7 %, general merchandise, same-store sales had fallen nearly 8 % the previous quarter.[186,187] CVS blamed that decline on the tobacco ban.[188]

To evaluate the public health impact of their decision to halt tobacco sales, the CVS Health Research Institute conducted a study the following year.[189] The study compared cigarette pack sales in states where CVS had a 15 % or greater share of the retail pharmacy market to states where its market share was less significant; it found an additional one percent decrease in cigarette sales in states where CVS had greater market share. The study also showed a 4 % increase in nicotine patch purchases in the states with greater CVS market share, signaling an increased consumer desire to quit smoking.[190] Though the results were modest, the decision to remove cigarettes from the shelves seemed to positively impact consumer health.

---

[183] CVS Health, "Our New Name," *CVS Health website,* September 3, 2014, http://www.cvshealth.com/newsroom/our-new-name, accessed December 2015.

[184] Hiroko Tabuchi, "How CVS Quit Smoking and Grew Into a Health Care Giant," *The New York Times,* July 11, 2015, http://www.nytimes.com/2015/07/12/business/how-cvs-quit-smoking-and-grew-into-a-health-care-giant.html, accessed January 2016.

[185] CVS Health, "CVS Health Reports Strong Profit Growth For Full Year 2014; Fourth Quarter Adjusted EPS At High End Of Company's Expectations," *CVS website,* February 10, 2015, accessed at http://www.cvshealth.com/content/cvs-health-reports-strong-profit-growth-full-year-2014-fourth-quarter-adjusted-eps-high-end, accessed January 2016.

[186] CVS Health, "CVS Health Reports Record Second Quarter Results," *CVS website,* August 4, 2015, accessed at http://www.cvshealth.com/content/cvs-health-reports-record-second-quarter-results, accessed January 2016.

[187] Matt Egan, "CVS banned tobacco. Now its sales are hurting," *CNN Money,* August 4, 2015, http://money.cnn.com/2015/08/04/investing/cvs-earnings-cigarettes/, accessed December 2015.

[188] Ibid.

[189] CVS Health, "CVS Health Marks First Anniversary of Tobacco Removal With New Data on Decision's Impact, Extends Commitment to Creating Tobacco-Free Generation," *Press release: CVS Health website,* September 3, 2015, https://www.cvshealth.com/content/cvs-health-marks-first-anniversary-tobacco-removal-new-data-decision%E2%80%99s-impact-extends, accessed December 2015.

[190] Ibid.

Conclusion

Through their products and services, businesses had profound impacts on consumer health. This case has assessed the reasons why some corporations took efforts to improve consumer health, as well as the methods by which they did so. There were many consumer health questions that remained.

Discussion Questions on Consumer Health

1. What level of responsibility should corporations have for providing consumer safety? Given that consumers have, and often want, freedom of choice, how much of this responsibility should fall to consumers? How much responsibility should fall to governments or other regulators?
2. To what degree should corporations or trade organizations influence policy and regulation standards? Are they overly influential today?
3. How should corporations forecast unintended public health consequences of new technologies (e.g., ear damage or hearing loss from headphones, carpal tunnel syndrome from keyboards)?

Chapter 3
Employee Health

On June 15th, 2015, *CNNMoney* published a short article titled "Amazon is America's best company. Says who? You!" According to the Reputation Institute, a firm that assesses public opinion on companies and industries, Amazon had the highest rating of any company in the US. The Reputation Institution stated that they determined ratings by measuring "a company's ability to deliver expectations in its products and services, innovation, workplace, governance, citizenship, leadership and financial performance."[1] By this account, it seemed that Amazon was not only meeting the demands of consumers, but surpassing other companies in their efforts.

Two months later, the *New York Times* featured an article on Amazon's workplace culture, depicting it as competitive, aggressive, and unforgiving for their employees. The article stated:

> Many of the newcomers filing in on Mondays may not be there in a few years. The company's winners dream up innovations that they roll out to a quarter-billion customers and accrue small fortunes in soaring stock. Losers leave or are fired in annual cullings of the staff—"purposeful Darwinism," one former Amazon human resources director said. Some workers who suffered from cancer, miscarriages and other personal crises said they had been evaluated unfairly or edged out rather than given time to recover.[2]

The article described how coworkers would often see one another cry and some employees struggled to remain at the employer due to high stress levels. The *New York Times* published a second article showcasing a handful of the comments and

---

Reprinted with permission of Harvard Business School Publishing
Employee Health, HBS No. 9-516-074
This case was prepared by John A. Quelch and Emily C. Boudreau.

[1] Aaron Smith, "Amazon is America's best company. Says who? You!," *CNN Money*, June 15, 2015, http://money.cnn.com/2015/06/15/news/companies/amazon-reputation/index.html, accessed October, 2015.

[2] Jodi Kantor and David Streitfeld, "Inside Amazon: Wrestling Big Ideas In a Bruising Workplace," *The New York Times*, August 15, 2015, http://www.nytimes.com/2015/08/16/technology/inside-amazon-wrestling-big-ideas-in-a-bruising-workplace.html, accessed October 2015.

J.A. Quelch, E.C. Boudreau, *Building a Culture of Health*, SpringerBriefs in Public Health, DOI 10.1007/978-3-319-43723-1_3

responses they received, some from Amazon employees themselves.[3] Some commenters reinforced the view the idea that Amazon had created an overwhelmingly stressful workplace, while others rushed to defend the technology company, stating that the benefits of working with thoughtful, driven coworkers on cutting-edge technology issues far outweighed the drawbacks of a fast-paced environment.[4]

In 2014, *Harvard Business Review*'s CEO ranking was based entirely on financial performance measures, and Jeff Bezos, Amazon's founder and CEO, ranked first. In 2015, Mr. Bezos ranked 87th.[5] The ranking system had changed in 2015, with 80 % of a CEO's score based on the traditional financial performance measures and 20 % based on a company's environmental, social, and governance (ESG) impact. *Harvard Business Review* stated, "On the purely financial metrics, Amazon's Jeff Bezos leads all other CEOs—just as he did last year. But Amazon's relatively poor ESG score drags Bezos down to #87 overall."[6]

While the exact nature of the Amazon workplace remained in dispute, the situation highlighted several key challenges that many other employers and their employees faced more generally. If Amazon, an innovative company that provided its employees with competitive salaries, benefits, and interesting opportunities within the technology industry could face these questions, certainly no company was immune to employee health and wellness challenges. Indeed, employee health was more frequently considered a matter of wellbeing, not simply the absence of disease.[7]

Could companies create fast-paced environments that drove revenue growth and innovative consumer offerings while simultaneously promoting physical and mental wellbeing among their employees? Was employee health an employer responsibility? Did corporate investments in employee health generate economic and/or social returns? How could corporations create high impact wellness programs? Were the returns on investments in employee health seen in the short or long-term?

Context

In 2014, full-time employed American adults spent an average of 47 h per week working and 18 % of adult workers reported working more than 60 h per week.[8] As a result, the nature of an employee's work, the health benefits available through their

[3] The New York Times, "Depiction of Amazon Stirs a Debate About Work Culture," *The New York Times*, August 18, 2015, http://www.nytimes.com/2015/08/19/technology/amazon-workplace-reactions-comments.html, accessed October 2015.

[4] Ibid.

[5] Harvard Business Review Staff, "The Best-Performing CEOs in the World," *Harvard Business Review November 2015 issue* (pp. 49–59), https://hbr.org/2015/11/the-best-performing-ceos-in-the-world, accessed October, 2015.

[6] Ibid.

[7] TE Kottke, M Stiefel, NP Pronk, "'Well-Being in All Policies': Promoting Cross-Sectoral Collaboration to Improve People's Lives," Prev Chronic Dis 2016;13:160155. DOI: http://dx.doi.org/10.5888/pcd13.160155.

[8] Lydia Saad, "The '40-Hour' Workweek Is Actually Longer—by Seven Hours," Gallup Poll website post, August 29, 2014, http://www.gallup.com/poll/175286/hour-workweek-actually-longer-seven-hours.aspx, accessed October 2015.

employer, and the level of stress their work created in both their professional and personal lives had significant impacts on their health.

While investment in employee health was not a new concept, there had been a rise in innovative wellness programs and increased debate around the appropriate role of employers in keeping employees well. Though the incentives for investing in employee health were significant, there were also many reasons why employers hesitated to make investments in employee wellness. Many programs aimed at improving employee wellness were resource intensive and required significant implementation time. This note will assess the drivers behind employer investments in health and wellness—as well as the benefits and tradeoffs of these investments.

Why Employers Invest in Employee Health

In 2012, approximately half of the US adult population had at least one chronic health condition (e.g., cancer, diabetes, obesity, and heart disease). Even more alarming, it appeared that chronic diseases were increasing worldwide, largely due to poor diet and inactivity.[9] (See Exhibit 3.1 for the increase in the percentage of US adults who had two or more chronic conditions between two time periods (1999–2000 vs. 2009–2010)).

Exhibit 3.1: US Adults with Two or More Chronic Conditions

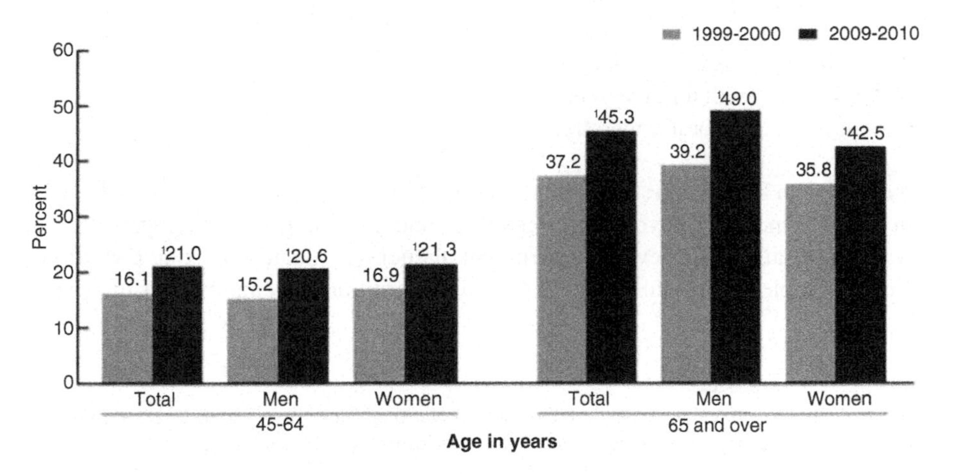

[9] World Health Organization, "Nutrition: Background," *WHO website*, http://www.who.int/nutrition/topics/2_background/en/, accessed November 2015.

Source: Virginia M. Freid, M.S.; Amy B. Bernstein, Sc.D.; and Mary Ann Bush, M.S., "Multiple Chronic Conditions Among Adults Aged 45 and Over: Trends Over the Past 10 Years," *CDC*: *NCHS DATA Brief*, No. 100, July 2012, accessed at http://www.cdc.gov/nchs/data/databriefs/db100.htm, accessed January 2016.

There was growing concern that the increasing prevalence of these conditions threatened the economic competitiveness of the US. In 2014, The Vitality Institute Commission on Health Promotion and the Prevention of Chronic Disease in Working-Age Americans issued five recommendations for combatting the increasing prevalence of chronic conditions: invest in prevention science; strengthen and expand leadership to deliver a unified message for health and prevention; make markets work for health promotion and prevention; integrate health metrics into corporate reporting; promote strong cross-sector collaborations to generate a systemic increase in health promotion and prevention across society.[10]

Physical health was not the only challenge—mental health issues, such as depression, were also a concern. At some point in their lives, 25 % of Americans would have a mental health issue, and 1 in 17 suffered from a more serious illness like bipolar disorder.[11,12]

By 2016, six key drivers had strengthened the importance of employer investments in combatting these significant health challenges many of their employees faced. The drivers were as follows:

1. Adhering to legislation
2. Managing costs
3. Generating revenue
4. Promoting brand reputation
5. Complying with union agreements
6. Addressing a moral imperative

Adhering to legislation: Though regulations varied substantially across the world, in the US, most employers were legally bound to adhere to certain standards of workplace safety, and several government agencies evaluated working conditions. Certain employee benefits, such as Social Security and unemployment insurance,

[10] The Vitality Institute, "Investing in Prevention: A National Imperative," June, 2014, http://thevitalityinstitute.org/site/wp-content/uploads/2014/06/Vitality_Recommendations2014.pdf, accessed May 2016.

[11] NIMH, "The Numbers Count: Mental Disorders in America," http://www.nimh.nih.gov/health/publications/the-numbers-count-mental-disorders-in-america/index.shtml#Intro, accessed July 2014.

[12] John A. Quelch and Carin-Isabel Knoop, "Mental Health and the American Workplace" HBS No. 9-515-062 (Boston: Harvard Business School Publishing, 2015), p. 2.

were mandated. Furthermore, when it was passed in 2010, the Patient Protection and Affordable Care Act (ACA) placed a new nationwide focus on healthcare in the US. Many of its provisions were directly aimed at providing health insurance to previously uninsured individuals, reducing the cost of care, and improving the quality of care. While these affected the types of health plans employers could offer their employees, the ACA had fewer direct impacts on employers than it did on previously uninsured individuals, health insurance providers, and healthcare providers.

In 2014, one study analyzed six separate microsimulation models that different researchers and organizations had created to assess the overall impact of the ACA. Across all of the models, the estimated change in covered lives ranged from a reduction of 6 million lives to an increase of 8 million lives.[13] The study author concluded that "though there are notable differences, all published microsimulation models suggest the overall effects of the Affordable Care Act on [employer-sponsored insurance] coverage will be modest."[14]

Despite the overall modest impact, several provisions did affect employers directly. One such provision was the Employer Shared Responsibility Provision, which penalized employers with 50 or more full time workers who did not offer health coverage to their full-time employees.[15] (See Exhibit 3.2 for a detailed overview of the Employer Shared Responsibility Provision.) The ACA also included additional incentives for "health-contingent" wellness programs.[16] Health-contingent wellness programs provided rewards based on health outcomes, often meaning that employees who met certain health criteria received a premium discount or other incentive. The ACA set the maximum reward at 30 % of the total cost of health coverage for most programs (shared between the employer and employee) and stipulated a higher reward amount, 50 % of coverage, for tobacco-related programs.[17]

---

[13] Fredric Blavin, Bowen Garrett, Linda Blumberg, Matthew Buettgens, Sarah Gadsden and Shanna Rifkin, "Monitoring the Impact of the Affordable Care Act on Employers," *The Urban Institute*, October 2014, http://www.urban.org/sites/default/files/alfresco/publication-pdfs/413273-Monitoring-the-Impact-of-the-Affordable-Care-Act-on-Employers.PDF, accessed October, 2015.

[14] Ibid.

[15] Kaiser Family Foundation, "Employer Responsibility Under the Affordable Care Act," October, 2, 2015, http://kff.org/infographic/employer-responsibility-under-the-affordable-care-act/, accessed October 2015.

[16] Karen Pollitz and Matthew Rae, "Workplace Wellness Programs Characteristics and Requirements," *The Kaiser Family Foundation*, June 15, 2015, http://kff.org/private-insurance/issue-brief/workplace-wellness-programs-characteristics-and-requirements/, accessed October 2015.

[17] Ibid.

Exhibit 3.2: Penalties for Employers Not Offering Coverage Under the Affordable Care Act During 2016

PENALTIES FOR EMPLOYERS NOT OFFERING COVERAGE UNDER THE AFFORDABLE CARE ACT DURING 2016

UPDATED AS OF OCTOBER 2, 2015

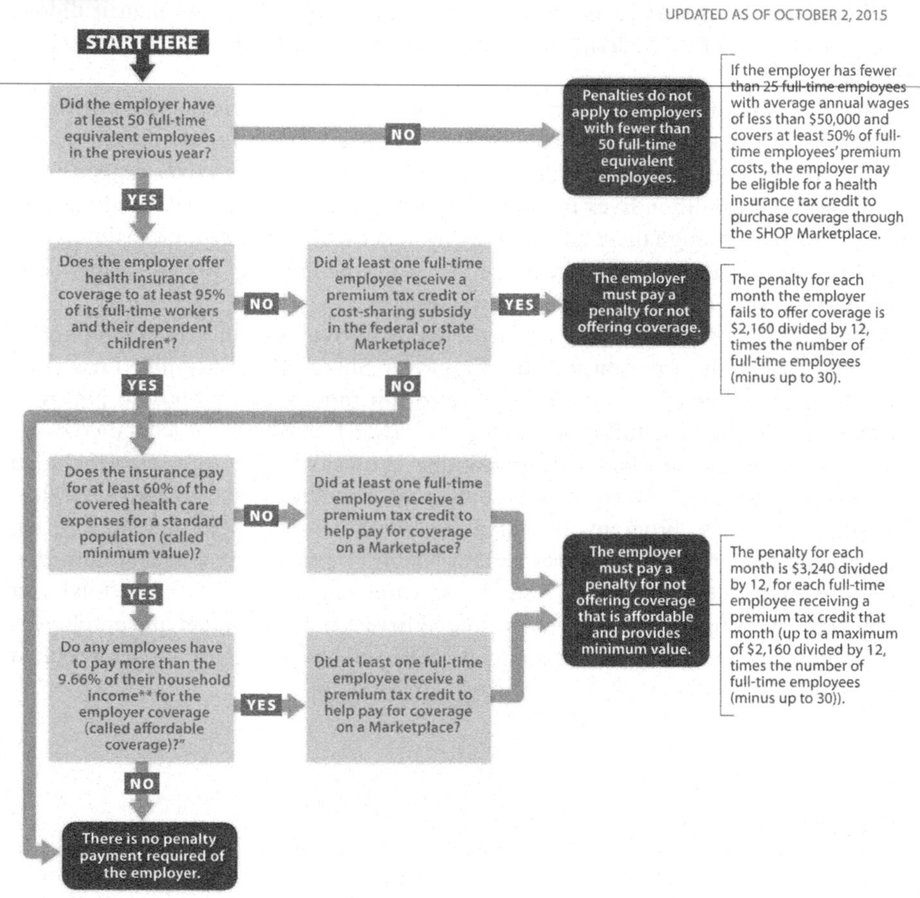

*A dependent child is defined as a child of an employee who is under the age of 26. Employers do not face a penalty under the Affordable Care Act if they do not offer coverage to the spouse of a full-time employee.

**Affordability is determined by reference to the taxpayer's household income (i.e., the employee's required contribution for self-only coverage should not exceed 9.66% in 2016). Since employers generally do not know the worker's household income, to determine if an employer may be subject to a penalty, the employer can measure 9.5% against three separate safe harbor amounts (the worker's Form W-2 wages, the worker's hourly rate of pay as of the first day of the coverage period, or the federal poverty line for a single individual). Note that a worker's and the worker's dependents' eligibility for premium tax credits or costing sharing subsidies are based on family income; not the safe harbor amounts.

http://kff.org/infographic/employer-responsibility-under-the-affordable-care-act/

Source: Kaiser Family Foundation, "Employer Responsibility Under the Affordable Care Act," *Kaiser Family Foundation*, October, 2, 2015, http://kff.org/infographic/employer-responsibility-under-the-affordable-care-act/, accessed October, 2015.

Managing costs: Employers had invested heavily in employee health to mitigate productivity costs associated with ill employees. These costs included lost productivity due to employee absenteeism, or "the time an employee spends away from work. Absences [could be] scheduled (e.g., vacation time) or unscheduled (e.g., due to illness or injury)."[18] Possibly as equally significant were the productivity costs associated with presenteeism, the extent to which health issues negatively affected employees who remained present in the workplace (e.g., an employee contracted the flu, but decided to attend work and was less productive than he/she was normally).[19] If an employee was absent or not working to his or her full capacity due to an illness, the employer had to use other resources to cover that loss.[20] US corporations lost around $225 billion annually due to absenteeism and presenteeism.[21]

Both mental and physical conditions contributed to these losses. For example, a study on the employer burden of major depressive disorder (MDD) in employees showed that MDD severity was not only associated with increased treatment usage/costs, but that there was also "a significant association between MDD severity and adjusted average monthly hours worked, work performance self-rating, the probability to miss one or more work days in past month, and monthly reduced work and performance."[22]

Compared with employers in countries like the UK, where the National Health Service (NHS) was paid for with tax dollars and supplied by the government, many employers in the US provided employer-sponsored health insurance as a benefit to aid employees in paying for their care. For these employers, cost management strategies were also aimed at reducing enrolled employees' health care utilization costs. Typically, insurers evaluated health care utilization costs across an employer's entire employee population on a routine basis, offering new premiums that appropriately captured the risk of that employee population.[23] This meant that "all employees within the workplace (or at least within broad job categories) typically [paid] the same amount for a health plan of given benefits and payment provisions," known as

---

[18] CDC, "Workplace Health Promotion: Glossary Terms," http://www.cdc.gov/workplacehealthpromotion/glossary/index.html#A2, accessed October 2015.

[19] Ibid.

[20] Thomas Parry and Bruce Sherman, "Workforce Health—The Transition from Costs to Outcomes to Business Performance," *Benefits Quarterly first quarter 2015*, http://www.ifebp.org/inforequest/ifebp/0166489.pdf, accessed October 2015.

[21] Bruce Japsen, "U.S. Workforce Illness Costs $576B Annually From Sick Days To Workers Compensation," *Forbes*, September 12, 2012, http://www.forbes.com/sites/brucejapsen/2012/09/12/u-s-workforce-illness-costs-576b-annually-from-sick-days-to-workers-compensation/#3ea7d8cf7256, accessed May 2016.

[22] Howard G. Birnbaum, et al., "Employer Burden of Mild, Moderate, and Severe Major Depressive Disorder: Mental Health Services Utilization and Costs, and Work Performance," Depression and Anxiety 27: 78–89 (2010).

[23] Thomas C. Buchmueller and Alan C. Monheit, "EMPLOYER-SPONSORED HEALTH INSURANCE AND THE PROMISE OF HEALTH INSURANCE REFORM," *The National Bureau of Economic Research*, April 2009, http://www.nber.org/papers/w14839.pdf, accessed October 2015.

a "community rate."[24] Thus, when employees, on average, accessed pricier care, employers tended to see an increase in health premiums. Keeping employees well had the potential to reduce health care utilization and premium increases over time.

Health insurance premium increases had skyrocketed throughout the 2000s, and employers were actively seeking ways to stem the trajectory.[25] One report showed that in the time period from 2010 to 2015, average premiums for covered workers with family coverage increased by about 27 %, surpassing both overall inflation and workers' earnings, at 9 % and 10 % respectively.[26] (See Exhibit 3.3 for health insurance premium increases compared to inflation and workers' earnings in three distinct time periods.)

Exhibit 3.3: Average Premium Increases for Covered Workers with Family Coverage, 2000–2015

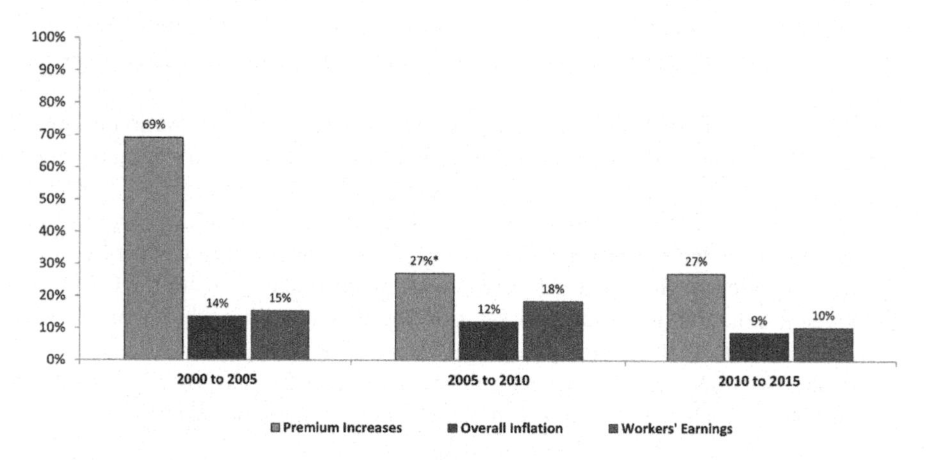

* Premium change is statistically different from previous period shown (p<.05).

SOURCE: Kaiser/HRET Survey of Employer-Sponsored Health Benefits, 2000-2015. Bureau of Labor Statistics, Consumer Price Index, U.S. City Average of Annual Inflation (April to April), 2000-2015; Bureau of Labor Statistics, Seasonally Adjusted Data from the Current Employment Statistics Survey, 2000-2015 (April to April).

Source: The Kaiser Family Foundation and Health Research & Educational Trust, "Employer Health Benefits: 2015 Summary Findings," September 22, 2015, http://kff.org/health-costs/report/2015-employer-health-benefits-survey/, accessed October, 2015.

[24] Ibid.

[25] The Kaiser Family Foundation and Health Research & Educational Trust, "Employer Health Benefits: 2015 Summary Findings," September 22, 2015, http://kff.org/health-costs/report/2015-employer-health-benefits-survey/, accessed October 2015.

[26] Ibid.

Generating revenue: Many employers started seeing the promotion of employee health as more than a cost management strategy. In the 2014 book *The Good Jobs Strategy*, Zeynep Ton argued that model retailers view their employees as assets and invest in them—and these retailers ultimately perform better financially.[27] Investing in employee health had the potential to improve company culture, ultimately increasing employee satisfaction and engagement—potentially increasing employee tenure.[28] Healthier, more engaged employees could think more creatively and drive innovative business ideas and revenue growth.

However, assessing the impact that a healthy, engaged employee could have on revenue growth potential proved more difficult than assessing the costs of an ill one. One article accurately portrayed this challenge by stating:

> Even the way the literature addresses health-related lost productivity is cost-based: the opportunity costs of lost work time, whether from absence or loss of performance. But research to date tells us little about top-line business performance impacts—how healthier employees may influence business results and how other health related factors the employer has influence over can contribute to business outcomes.[29]

Promoting brand reputation: Employers also invested in employee health to improve their company and brand reputation. Well-known company ranking systems, such as FORTUNE's 100 Best Companies to Work For®, depended heavily on employee satisfaction and opinion. Two-thirds of Fortune's ranking was based on "questions related to employees' attitudes about management's credibility, overall job satisfaction, and camaraderie," while the final third was based on "questions about pay and benefit programs and a series of open-ended questions about hiring practices, methods of internal communication, training, recognition programs, and diversity efforts."[30]

In *The Good Jobs Strategy*, Ton pointed out that online communities like Glassdoor made it simpler for employees to anonymously share their experiences with others—discussing how, in August, 2012, Trader Joe's scored an 84 % on the question, "Would you recommend this employer to a friend?," while Walmart scored a 47 %.[31] Therefore, investing in these areas had the potential to improve a company's ranking and visibility with potential employees and consumers.

---

[27] Zeynep Ton, "The Good Jobs Strategy," *Houghton Mifflin Harcourt Publishing Company*, New York, New York, 2014, pp. 64–67.

[28] Ibid.

[29] Thomas Parry and Bruce Sherman, "Workforce Health—The Transition from Costs to Outcomes to Business Performance," *Benefits Quarterly first quarter 2015*, http://www.ifebp.org/inforequest/ifebp/0166489.pdf, accessed October 2015.

[30] Fortune, "100 Best Companies to Work For," Fortune website, http://fortune.com/best-companies/, accessed October 2015.

[31] Zeynep Ton, "The Good Jobs Strategy," *Houghton Mifflin Harcourt Publishing Company*, New York, New York, 2014, p. 67.

Promoting employee health and wellness was also often part of a larger corporate social responsibility (CSR) strategy, which had gained traction and popularity among employers in the 1990s and 2000s.[32] Though the authors of one study on consumer decision-making cautioned that the more traditional drivers of consumer decision-making, price and quality, remained most impactful, they found a relationship between investments in CSR programs and improved consumer opinion, stating:

> Companies that evidence societal responsibility have been rewarded for their efforts and behaviors—positive word of mouth among consumers, stronger market position, and thus superior financial performance, as compared to companies with less responsible practices.[33]

Complying with union agreements: Labor unions collectively negotiated with employers on the behalf of employees around issues like working conditions, wages, and benefits. The Wagner Act of 1935 gave employees the right to form and join unions, and also required employers to work with them.[34] In 1983, the union membership rate among wage and salary workers was 20.1 %; however, the rate was down to 11.1 % by 2014.[35] Unions had a more significant role in the public sector, where 35.7 % of workers belonged to a union in 2014, than in the private sector, where only 6.6 % of the workers belonged to a union.[36] For industries that were heavily unionized, including teachers and police officers, union requests represented a final driver for employer investments in employee health.

Addressing a moral imperative: While there were clearly business reasons to make investments in employee health, many corporate leaders and their boards invested in employee health because they believed that, morally, it was the right thing to do. The fact was that work was a major cause of physical and mental stress—and sometimes illness—for many employees. Therefore, many morally-led companies felt a sense of duty to address employee health concerns—especially those that were clearly associated with the workplace, such as high stress levels.

---

[32] Elias G. Rizkallah, "Brand-Consumer Relationship And Corporate Social Responsibility: Myth Or Reality & Do Consumers Really Care?," *Journal of Business and Economics Research*, June 2012, Vol. 10, No. 6.

[33] Elias G. Rizkallah, "Brand-Consumer Relationship And Corporate Social Responsibility: Myth Or Reality & Do Consumers Really Care?," *Journal of Business and Economics Research*, June 2012, Vol. 10, No. 6.

[34] National Labor Relations Board, "The 1935 passage of the Wagner Act," https://www.nlrb.gov/who-we-are/our-history/1935-passage-wagner-act, accessed October 2015.

[35] Bureau of Labor Statistics, "Union Members Summary," Bureau of Labor Statistics Economic Summary, http://www.bls.gov/news.release/union2.nr0.htm, accessed October, 2015.

[36] Ibid.

How Employers Promote Employee Health

In October 2015, Danny Meyer, the chief executive of Union Square Hospitality Group ("USHG"), announced that about a dozen of his New York City restaurants would eliminate the practice of tipping and increase meal prices to compensate for the difference. This change was in an effort to raise the wages of dining room managers and kitchen staff, as they did not receive gratuities or share in pooled tips from wait staff.[37] Mr. Meyer said that over the last 30 years "the gap between what the kitchen and dining room workers make has grown by leaps and bounds…kitchen income has gone up no more than 25 %. Meanwhile, dining room pay has gone up 200 %."[38]

There was increasing evidence that promoting a positive workplace environment was critical for employee health and wellness. Combatting inequity, as the USHG did, was one strategy businesses took. Others included improving workplace safety standards, changing benefits, raising wage levels, or improving company culture. Many of these changes were particularly important for combatting mental health challenges in the workplace.

The main ways that employers promoted employee health were through changes in the following areas:

1. Workplace safety
2. Employee benefit programs
   - Health insurance
   - Workplace wellness programs
   - Mental health programs
   - Other company-specific benefits
3. Wage standards
4. Company structure and culture

Workplace Safety

Workplace safety was one of the most fundamental aspects of employer investments in employee health and it could contribute to both physical and mental wellness issues if not properly addressed. In 1884, Germany became the first country to

[37] Pete Wells, "Danny Meyer Restaurants to Eliminate Tipping," New York Times, October 14, 2015, http://www.nytimes.com/2015/10/15/dining/danny-meyer-restaurants-no-tips.html?_r=0.
[38] Ibid.

provide compensation to employees who were injured in workplace-related accidents, and many other countries followed suit not long after.[39] Workers' Compensation laid the groundwork for some of the first, significant employer investments in working conditions, as it gave employers a financial incentive to avoid work-related accidents. The Department of Labor explained the economic incentives of workers' compensation as:

> The economic incentive to employers was tied to the premiums they would pay to the state or to the insurance company that underwrote compensation payments. Premiums would vary from industry to industry and from plant to plant, depending on the safety record, extent of hazards, and the efforts companies made in accident prevention. The employer would pay a lower rate as safety efforts and the accident record improved in his company or his industry. Supporters of workmen's compensation asserted that the potential savings in premiums would more than cover the costs of necessary safety improvements, making it in the employers' own best economic interest to reduce accident rates.[40]

Occupational safety standards in the US were further advanced in 1970 when President Richard Nixon signed the Williams-Steiger Occupational Safety and Health Act into law. The Act established the Occupational Safety and Health Administration (OSHA), the National Institute of Occupational Safety and Health (NIOSH), and the independent Occupational Safety and Health Review Commission.[41]

While the organizations worked together, NIOSH sat within the U.S. Centers for Disease Control and Prevention and became largely a research organization, while OSHA was a part of the U.S. Department of Labor and took on a regulatory and enforcement role.[42] The Occupational Safety and Health Review Commission was an independent federal agency that reviewed employer appeals to OSHA decisions.[43]

Although significant advancements occurred across most industries, some occupations remained riskier than others. In 2014, the industries with the greatest number of fatal occupational injuries were construction, transportation and warehousing, and agriculture, forestry, fishing, and hunting. See Exhibit 3.4 for workplace fatalities by sector in 2014.

---

[39] United States Department of Labor, "Progressive Ideas," http://www.dol.gov/dol/aboutdol/history/mono-regsafepart06.htm, accessed October 2015.

[40] Ibid.

[41] OSHA, "OSHA Celebrates 40 years of accomplishments in the Workplace," https://www.osha.gov/osha40/OSHATimeline.pdf, accessed October 2015.

[42] The National Institute for Occupational Safety and Health (NIOSH), "About NIOSH," http://www.cdc.gov/niosh/about.html, accessed October 2015.

[43] Occupational Safety and Health Review Commission, "About the Commission," http://www.oshrc.gov/, accessed October 2015.

Exhibit 3.4: Number and Rate of Fatal Occupational Injuries by Industry Sector, 2014

Number and rate of fatal occupational injuries by industry sector, 2014*

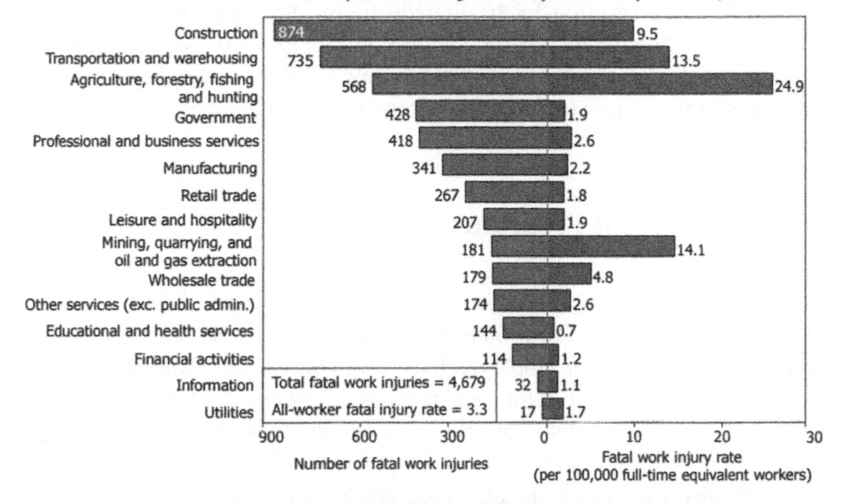

Private construction had the highest count of fatal injuries in 2014, but the private agriculture, forestry, fishing and hunting sector had the highest fatal work injury rate.

*Data for 2014 are preliminary.
Note: Fatal injury rates exclude workers under the age of 16 years, volunteers, and resident military. The number of fatal work injuries represents total published fatal injuries before the exclusions. For additional information on the fatal work injury rate methodology, please see http://www.bls.gov/iif/oshnotice10.htm.
Source: U.S. Bureau of Labor Statistics, Current Population Survey, Census of Fatal Occupational Injuries, 2015.

14

Source: Bureau of Labor Statistics, "Number and rate of fatal occupational injuries by industry sector, 2014*," http://www.bls.gov/iif/oshwc/cfoi/cfch0013.pdf, accessed October, 2015.

Despite the progression of workplace safety standards in the US, the April 2013 collapse of the Rana Plaza building in Bangladesh highlighted that workplace safety was still an important challenge to employee wellness globally. The building collapse resulted in the death of over 1100 factory employees and rescue workers.[44] The event caused international outcry, and in June 2015, Bangladesh charged 41 people involved with murder charges.[45]

Primark Stores Ltd. ("Primark"), a division of Associated British Foods (ABF) and an Irish clothing manufacturer, had a contract with one of the garment factories in the Rana Plaza building. It responded quickly to compensate the victims of the Rana Plaza tragedy. To fairly and appropriately distribute aid, Primark developed a three-part strategy that included an actuarial model, a medical assessment and a

---

[44] John A. Quelch and Margaret L. Rodriguez, "Rana Plaza (C): Primark and Victim Compensation" HBS No. 9-516-014 (Boston: Harvard Business School Publishing, 2015).

[45] JULFIKAR ALI MANIK and NIDA NAJAR, "Bangladesh Police Charge 41 With Murder Over Rana Plaza Collapse," *The New York Times*, June 1, 2015, http://www.nytimes.com/2015/06/02/world/asia/bangladesh-rana-plaza-murder-charges.html, accessed October, 2015.

vulnerability assessment.[46] In March of 2015, Primark publically announced that the company had distributed more than 95 % of the expected long-term compensation payments. At that time, Primark had contributed a total of $14 million to victims.[47] Primark's reaction to the tragedy highlighted the ongoing importance of employer investments in creating and promoting a safe working environment. The company's actions also showcased how taking socially conscious actions when that fails can restore employee faith and improve public relations.

Employee Benefit Programs

Employee benefit packages had become a staple component of total compensation. In 2015, the U.S. Bureau of Labor Statistics reported that private industry employers spent an average of $31.39 per employee for each hour worked, with wages and salaries accounting for 69.5 % of that total and benefits accounting for the remaining 30.5 %.[48]

Beyond legally mandated benefits, employers used employee benefit programs to attract and retain employees. For example, most Silicon Valley startup companies provided subsidized or free healthy food options for lunch.[49] Many millennials considering employment at such companies all but demanded this perk. One *Business Insider* article stated, "Free snacks, yoga class, bottomless drinks, and a back massage may sound like a night at an all-inclusive luxury resort, but for employees at Pinterest it's just another day at the office. Tech companies desperate for talent are engaged in an all out free perks arms race…"[50]

While there were some employee benefits that were more directly linked to employee health (e.g., health insurance and wellness programs that promoted physical and mental health), there were many more that were tied indirectly. These included retirement, paid leave (maternity, paternity, illness), financial incentives (e.g., stock options, bonuses), and flexible work arrangements.

[46] John A. Quelch and Margaret L. Rodriguez, "Rana Plaza (C): Primark and Victim Compensation" HBS No. 9-516-014 (Boston: Harvard Business School Publishing, 2015).

[47] Primark, "Rana Plaza," *Primark website*, http://www.primark.com/en/our-ethics/news/rana-plaza, accessed October, 2015.

[48] Bureau of Labor Statistics, "EMPLOYER COSTS FOR EMPLOYEE COMPENSATION—JUNE 2015," News Release, September 9, 2015, http://www.bls.gov/news.release/pdf/ecec.pdf, accessed October, 2015.

[49] Taylor Lorenz, "Startups Offer So Many Extravagant Perks, They Have Begun Hiring People Just To Manage It All," November, 25, 2014, http://www.businessinsider.com/startups-offer-so-many-extravagant-perks-they-have-begun-hiring-people-just-to-manage-it-all-2014-11, accessed May, 2016.

[50] Ibid.

Google Inc.'s ("Google") investment in mental wellbeing showcased the way that many companies were placing a newfound focus on innovative benefits that improved employee wellness in less traditional ways. The company provided several avenues for employees to assess and address their stress. Google's Search Inside Yourself program focused on mindfulness to develop resilience, acknowledging that their employees' work could be stressful, but that there were healthful ways to address that stress.[51]

Chade-Meng Tan was Google's "Jolly Good Fellow," and his job was to foster emotional wellbeing and happiness among Google employees. Although he started as an early engineer at Google, Mr. Tan later created and taught a class at Google on mindfulness.[52] Two thousand employees had taken his class, and Mr. Tan believed that mindfulness led to increased kindness, and ultimately, better business practices.[53] He said:

> In many situations, goodness is good for business...If you, as the boss, are nice to your employees, they are happy, they treat their customers well, the customers are happy to spend more money, so everybody wins. Also if you treat everybody with kindness, they'll like you even if they don't really know why. And if they like you, they want to help you succeed. So it's good for your soul and it's good for your career.[54]

Health Insurance

One of the employee benefits most directly linked to health was employer-sponsored health insurance. In 2015, over half of the non-elderly population in the United States was covered by employer-sponsored health insurance.[55]

As previously noted, health insurance premiums had increased significantly across the 2000s.[56] In employer-sponsored plans, premiums were partially funded by the employer and partially funded by the employee. The employee share of the premium had remained relatively stable for the prior two decades, meaning that

---

[51] Drake Baer, "Here's What Google Teaches Employees In Its 'Search Inside Yourself' Course," *Business Insider*, August 5, 2014, http://www.businessinsider.com/search-inside-yourself-googles-life-changing-mindfulness-course-2014-8, accessed May, 2016.

[52] Julie Bort, "This Google engineer's title is 'Jolly Good Fellow' and he's solving unhappiness and war," *Business Insider*, September 18, 2015. http://www.businessinsider.com/google-jolly-good-fellow-chade-meng-tan-2015-9.

[53] The Guardian, "Google's head of mindfulness: 'goodness is good for business'," http://www.theguardian.com/sustainable-business/google-meditation-mindfulness-technology, accessed October 2015.

[54] Ibid.

[55] The Kaiser Family Foundation and Health Research & Educational Trust, "Employer Health Benefits: 2015 Summary Findings," September 22, 2015, http://kff.org/health-costs/report/2015-employer-health-benefits-survey/, accessed October 2015.

[56] The Kaiser Family Foundation and Health Research & Educational Trust, "Employer Health Benefits: 2015 Summary Findings," September 22, 2015, http://kff.org/health-costs/report/2015-employer-health-benefits-survey/, accessed October 2015.

premium costs were increasing for both the employer and employee.[57] However, Katherine Baicker and Amitabh Chandra noted that "increases in health insurance premiums do not get absorbed by an unlimited reservoir of profits or endowments—they are paid for by employees taking home smaller paychecks."[58] This suggested that premium increases were ultimately financed by the employee.

To control the increases in premiums and encourage employees to take more active roles in their own care, employers and health insurance providers had used cost-shifting mechanisms. This trend was most abundantly clear through the proliferation of high-deductible health plans (HDHPs).[59] (See Exhibit 3.5 for trends in HDHPs.) HDHPs had lower monthly premiums than more traditional plans.[60] Employees typically still received catastrophic coverage in the event of a serious illness and lower, negotiated prices due to their participation in the plan. However, because the plans offered lowered premiums, they came with higher deductibles that employees had to meet through out-of-pocket spending.[61]

Exhibit 3.5: Percentage of Covered Workers Enrolled in an HDHP/HRA or HSA-Qualified HDHP, 2006–2015

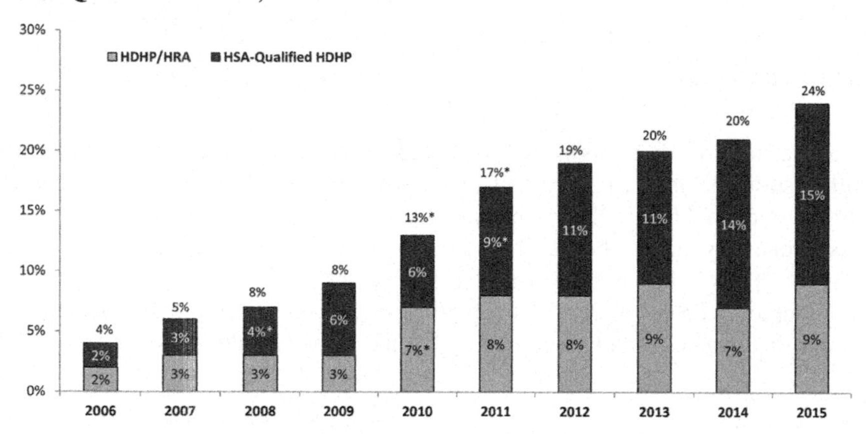

[57] Katherine Baicker and Amitabh Chandra, "The Veiled Economics of Employee Cost Sharing," *JAMA Internal Medicine*, May 2015, Vol. 175, no. 7, pp. 1081–2.

[58] Katherine Baicker and Amitabh Chandra, "The Veiled Economics of Employee Cost Sharing," *JAMA Internal Medicine*, July 2015, Vol. 175, no. 7, pp. 1081–2.

[59] The Kaiser Family Foundation and Health Research & Educational Trust, "Employer Health Benefits: 2015 Summary Findings," September 22, 2015, http://kff.org/health-costs/report/2015-employer-health-benefits-survey/, accessed October 2015.

[60] Ibid.

[61] Ibid.

Source: The Kaiser Family Foundation and Health Research & Educational Trust, "Employer Health Benefits: 2015 Summary Findings," September 22, 2015, http://kff.org/health-costs/report/2015-employer-health-benefits-survey/, accessed October, 2015.

Did this switch empower employees to price shop and become savvier consumers of care, or did it cause them to avoid care in an effort to reduce out-of-pocket costs? In a 2015 study published by the National Bureau of Economic Research (NBER), researchers found that when employees were required to switch from an insurance plan that provided free health care to an HDHP, total healthcare spending went down—due entirely due to reductions in the quantity of care.[62] These reductions were seen for both "potentially valuable care (e.g. preventive services) and potentially wasteful care (e.g. imaging services)."[63] Interestingly, the researchers found no evidence that employees were price shopping for care within two years of switching to the HDHP coverage.[64]

Cost-sharing mechanisms, such as increased deductibles and copayments, seem to have had both benefits and drawbacks.[65] They might have reduced the overuse of medical care and reduce further premium increases, but they also had the potential to dissuade an individual from accessing necessary care.[66] In 2015, Baicker and Chandra suggested that a more nuanced understanding and use of cost-sharing mechanisms (e.g., how cost-sharing mechanisms affect different kinds of care and patients with different income) was necessary to "improve health, slow increases in health insurance premiums, and increase take-home pay."[67]

The ACA required employers with more than 50 full-time employees to provide health insurance to employees working more than 30 h a week.[68] Part-time workers who did not obtain coverage through their employment were able to purchase health insurance on the Health Insurance Marketplace. In 2014, Walmart raised premiums for all employees and cut health insurance benefits for 30,000 part-time employees

[62] Zarek C. Brot-Goldberg, Amitabh Chandra, Benjamin R. Handel, Jonathan T. Kolstad, "What Does a Deductible Do? The Impact of Cost-Sharing on Health Care Prices, Quantities, and Spending Dynamics," *The National Bureau of Economic Research*, October 2015, NBER Working Paper No. 21632, accessed at http://www.nber.org/papers/w21632, accessed January 2016.

[63] Ibid.

[64] Ibid.

[65] Katherine Baicker and Amitabh Chandra, "The Veiled Economics of Employee Cost Sharing," *JAMA Internal Medicine*, May 2015, Vol. 175, no. 7, pp. 1081–2.

[66] Ibid.

[67] Ibid.

[68] IRS, "Questions and Answers on Employer Shared Responsibility Provisions Under the Affordable Care Act," https://www.irs.gov/Affordable-Care-Act/Employers/Questions-and-Answers-on-Employer-Shared-Responsibility-Provisions-Under-the-Affordable-Care-Act, accessed October 2015.

who worked less than 30 h a week.[69] This group made up about 2–5 % of the company's part-time workforce at the time.[70,71] Walmart defended their decision with a blog post on their website. It stated:

> Like every company, Walmart continues to face rising health care costs. This year, the expenses were significant and led us to make some tough decisions as we begin our annual enrollment. As a result, today we announced that our associates will see an increase in premiums for 2015. For example, our most popular and lowest cost associate-only plan will increase by $3.50 to $21.90 per pay period—still half the average premium other retail employees pay.
>
> We're also changing eligibility for some part-time associates. We will continue to provide affordable health care to all eligible associates, including part-time, who work more than 30 hours. However, similar to other retailers like Target, Home Depot, Walgreens and Trader Joe's, we will no longer be providing health benefits to part-time associates who work less than 30 hours.[72]

Workplace Wellness Programs

In 2015, more than 80 % of US employers offered an employee wellness program as a benefit to help employees lose weight, improve chronic health conditions, stop smoking or make other healthy lifestyle modifications.[73] Though definitions of workplace wellness programs have varied, in 2012, a comprehensive review of workplace wellness programs defined a workplace wellness program with the following statement:

> Broadly, a workplace wellness program is an employment-based activity or employer sponsored benefit aimed at promoting health-related behaviors (primary prevention or health promotion) and disease management (secondary prevention). It may include a combination of data collection on employee health risks and population-based strategies paired with individually focused interventions to reduce those risks. A formal and universally accepted

[69] HIROKO TABUCHI, "Walmart to End Health Coverage for 30,000 Part-Time Workers," *The New York Times*, October 7, 2014, http://www.nytimes.com/2014/10/08/business/30000-lose-health-care-coverage-at-walmart.html, accessed October 2015.

[70] Sally Welborn, "Providing Quality Health Benefits for Our Associates," *Walmart Today* blog, Walmart website, http://blog.walmart.com/business/20141007/providing-quality-health-benefits-for-our-associates, accessed October 2015.

[71] HIROKO TABUCHI, "Walmart to End Health Coverage for 30,000 Part-Time Workers," *The New York Times*, October 7, 2014, http://www.nytimes.com/2014/10/08/business/30000-lose-health-care-coverage-at-walmart.html, accessed October 2015.

[72] Sally Welborn, "Providing Quality Health Benefits for Our Associates," *Walmart Today* blog, Walmart website, http://blog.walmart.com/business/20141007/providing-quality-health-benefits-for-our-associates, accessed October 2015.

[73] The Kaiser Family Foundation and Health Research & Educational Trust, "Employer Health Benefits: 2015 Summary Findings," September 22, 2015, http://kff.org/health-costs/report/2015-employer-health-benefits-survey/, accessed October 2015.

definition of a workplace wellness program has yet to emerge, and employers define and manage their programs differently.[74]

Some employers created their own programs, but many used third-party wellness vendors. By 2014, workplace wellness was a $6 billion dollar industry in the US.[75] The industry had grown rapidly, and some of the programs had become highly personalized. For example, Newtopia, a genetics-based wellness program offered genetic testing followed by personalized coaching for disease prevention.[76]

Programs varied in their goals, scope, cost, and use of incentives. Incentives were used to encourage eligible employees to take part in available workplace wellness programs. In 2014, one study found that the employee participation rate for employers that did not use incentives was around 20 %, while employers who used rewards had a 40 % participation rate, and employers who used penalties and/or rewards had a 73 % participation rate.[77]

In 2015, The Institute for Health and Productivity Studies at the Johns Hopkins Bloomberg School of Public Health collaborated with the Transamerica Center for Health Studies to explore, among other things, which incentives were most likely to change employees' behaviors in a cost effective way. Although many programs provided some type of incentive, its findings indicated that it was important to provide positive, visible rewards related to healthcare; what's more, some companies opted to give rewards smaller rewards over time to sustain employee engagement.[78]

Questions remained about the future wellness program incentives, as there had been several lawsuits over their use and there were several more pending in early 2016.[79] In one case, a corporation denied health insurance coverage to a worker who

---

[74] Soeren Mattke, Christopher Schnyer, Kristin R. Van Busum, "A Review of the U.S. Workplace Wellness Market, "*RAND CORPORATION*, 2012, p. 5, accessed at http://www.dol.gov/ebsa/pdf/workplacewellnessmarketreview2012.pdf, accessed October 2015.

[75] RAND Corporation, "Do Workplace Wellness Programs Save Money?" 2014, http://www.rand.org/content/dam/rand/pubs/research_briefs/RB9700/RB9744/RAND_RB9744.pdf, accessed October 2015.

[76] Rachel Emma Silverman, "Genetic Testing May Be Coming to Your Office," *The Wall Street Journal*, December 15, 2015, accessed at http://www.wsj.com/articles/genetic-testing-may-be-coming-to-your-office-1450227295, accessed January 2016.

[77] Soeren Mattke, Kandice Kapinos, John P. Caloyeras, Erin Audrey Taylor, Benjamin Batorsky, Hangsheng Liu, Kristin R. Van, Busum, Sydne Newberry, "Workplace Wellness Programs Services Offered, Participation, and Incentives," *RAND Corporation*, 2014, https://www.dol.gov/ebsa/pdf/WellnessStudyFinal.pdf, accessed October 2015.

[78] Hector De La TorreRon Goetzel, "From Evidence to Practice: Workplace Wellness that Works," *Harvard Business Review*, September, 2015, https://hbr.org/2016/03/how-to-design-a-corporate-wellness-plan-that-actually-works, accessed May 2016.

[79] Rebecca Greenfield, "Employee Wellness Programs Not So Voluntary Anymore," *Bloomberg Business*, January 15, 2016, accessed at http://www.bloomberg.com/news/articles/2016-01-15/employee-wellness-programs-not-so-voluntary-anymore?utm_source=hs_email&utm_medium=email&utm_content=25343526&_hsenc=p2ANqtzDOUBLEHYPHENh1TFnT8sM2-stfRLQyLr7SLjqDMfLdICnIETggN1OPz-LM91KNyVkTpGwMcg-L3njOKpGROhIBI-Lt1RhJl2WOUy_ZLQ&_hsmi=25343526, accessed January 2016.

refused to participate in the employer's work-sponsored health assessment and biometric screening.[80] In 2015, a federal court judge in Wisconsin ruled in favor of that employer—reasoning that employers could require health assessment participation and deny coverage so long as the data obtained was used for determining the overall risk of its insurance pool.[81]

Some companies that had heavily invested in employee health had seen significant returns. Johnson & Johnson (J&J), a medical device, pharmaceutical, and consumer packaged goods company, had made employee health and wellness a priority for decades. The company also evaluated many of its wellness investments, which cost over $60 million per year by 2014,[82] over time. In 2014, J&J's "Culture of Health" 12-program framework included a tobacco-free workplace, free health profiles, an employee assistance program, medical surveillance, physical activity, health promotion, stress and energy management, cancer awareness, HIV/AIDs awareness, healthy eating, modified duty/return to work, and travel health. All global locations participated, though program content was customized by location and culture. Health tools included prevention-focused education, rewards for healthy behaviors, and workplace environments that made it natural for employees to engage in healthy behaviors.[83]

J&J began to collect data on health-related savings from their investments in wellness programs as early as the 1990s. In 1995, medical claims for around 19,000 US-based employees who participated in the Health&Wellness program showed an average total savings of $224.66 per employee across all expenditure categories, with savings becoming more significant over time.[84] A separate study analyzed employee health risk and cost data from 2002 to 2008. Researchers found that the average J&J employee had a lower predicted probability of being at high risk for high blood pressure, high cholesterol, poor nutrition, obesity, physical inactivity, and tobacco use; however, J&J employees were at higher risk for depression and stress than employees in an external comparison group.[85] In the time period studied,

---

[80] Ibid.

[81] Ibid.

[82] John A. Quelch and Carin-Isabel Knoop, "Johnson & Johnson: The Promotion of Wellness," HBS No. 9-514-112 (Boston: Harvard Business School Publishing, 2014).

[83] John A. Quelch and Carin-Isabel Knoop, "Johnson & Johnson: The Promotion of Wellness," HBS No. 9-514-112 (Boston: Harvard Business School Publishing, 2014).

[84] Ronald J. Ozminkowski, et al., "Long-Term Impact of Johnson & Johnson's Health & Wellness Program on Health Care Utilization and Expenditures," *Journal of Occupational and Environmental Medicine*, Vol. 44, No. 1, 2002, http://www.thehealthproject.com/documents/2003/johnson_johnson_utilization_expenditure.pdf, accessed January 2014.

[85] Rachel M. Henke, Ron Z. Goetzel, Janice McHugh, and Fik Isaac, "Recent Experience in Health Promotion at Johnson & Johnson: Lower Health Spending, Strong Return On Investment," *Health Affairs 30:3*, March 2011, via ProQuest Business Collection, accessed January 2014.

2002–2008, J&J's medical spending grew at a 3.7 % lower annual rate than at comparable US companies.[86,87]

Despite successes like J&J and the widespread and increasing use of workplace wellness programs, many expressed concerns with the programs, which included:

1. Unknown or lack of impact
2. Inadequate protection of employee privacy
3. Creation of a toxic or competitive workplace culture

In 2012, one study assessed 33-peer reviewed publications on workplace wellness programs and found that only five provided ROI estimates. The ROI estimates were modest and ranged from $1.65 to $6 per dollar spent.[88] However, the study advised that these positive results should be viewed cautiously, as most programs were not evaluated scientifically rigorously and there was a possible bias towards reporting positive results.[89]

Privacy was also a chief concern, as employees supplied large amounts of healthcare data to wellness programs, and some feared that there were not enough protections for employee data. This created the potential that "it could be abused for workplace discrimination, credit screening or marketing."[90]

There were also concerns that workplace wellness programs could actually increase employee stress, thus backfiring and making employees less well. A group of researchers who conducted analyses on the culture and history of wellness in the workplace stated that this occurred as "employees ploughed a great deal of energy into trying to improve their health. Sometimes this meant that employees had less time to focus on their core tasks. More frequently, these wellness initiatives would eat into employees' personal lives."[91]

Further, they cautioned that organizations could become too heavily focused on health and fitness, creating cultures where relatively healthy people felt they could not fit in.[92] A 2015 article in *The Economist* noted how this cultural shift could be

---

[86] Rachel M. Henke, Ron Z. Goetzel, Janice McHugh, and Fik Isaac, "Recent Experience in Health Promotion at Johnson & Johnson: Lower Health Spending, Strong Return On Investment," *Health Affairs 30:3*, March 2011, via ProQuest Business Collection, accessed January 2014.

[87] John A. Quelch and Carin-Isabel Knoop, "Johnson & Johnson: The Promotion of Wellness," HBS No. 9-514-112 (Boston: Harvard Business School Publishing, 2014).

[88] Soeren Mattke, Christopher Schnyer, Kristin R. Van Busum, "A Review of the U.S. Workplace Wellness Market," *RAND Corporation*, 2012, http://www.dol.gov/ebsa/pdf/workplacewellnessmarketreview2012.pdf, accessed October 2015.

[89] Ibid.

[90] Jay Hancock and Julie Appleby, "*7 Questions To Ask Your Employer About Wellness Privacy*," *Kaiser Health News*, September, 30, 2015, http://khn.org/news/7-questions-to-ask-your-employer-about-wellness-privacy/, accessed October 2015.

[91] Scott Berinato, "Corporate Wellness Programs Make Us Unwell: An Interview with André Spicer," *Harvard Business Review May 2015 issue* (pp. 28–29), https://hbr.org/2015/05/corporate-wellness-programs-make-us-unwell, accessed October 2015.

[92] Ibid.

particularly difficult for CEOs and other organizational leaders, who might be under more pressure to perform than the average employee.[93] This had caused some to subscribe to intense fitness regimens or take medications that aided focus. The article stated, "it is time to call a halt on all this hyperactivity, before it gets out of hand. There is no doubt that many bosses have heavy weights resting on their shoulders. But are they likely to make these decisions better if they arrive at work exhausted and sleep-deprived?"[94]

Workplace wellness programs were continually evolving and many questions remained about their future directions. One reviewer discussed the wide array of unknowns in the industry, stating:

> ...there is little insight offered on the proportion of funds to be spent on individual counseling, biometric testing, exercise equipment, classes, subsidies for healthy food items, Web sites, advisory group meetings, personal trainers, on-site clinics, and incentives, to name but a few program investment alternatives. More research is, therefore, needed to better understand which of the above program components is most cost-effective in providing value to workers, driving participation, and achieving specific outcomes. In addition, research is needed to determine which program components are most applicable for certain populations. For example, best practices found in a manufacturing facility may not translate well to call centers, universities, or hospitals.[95]

Mental Health Programs

According to the World Health Organization (WHO), there was "increasing evidence that both the content and context of work can play a role in the development of mental health problems in the workplace."[96] For example, research showed that disengaged employees were more likely to have health issues (e.g., stress, high

---

[93] The Economist, "Here comes SuperBoss," *The Economist*: *Schumpeter*," December 19, 2015, http://www.economist.com/news/business/21684107-cult-extreme-physical-endurance-taking-root-among-executives-here-comes-superboss, accessed January 2016.

[94] Ibid.

[95] RZ Goetzel, Henke RM, Tabrizi M, Pelletier KR, Loeppke R, Ballard DW, Grossmeier J, Anderson DR, Yach D, Kelly RK, McCalister T, Serxner S, Selecky C, Shallenberger LG, Fries JF, Baase C, Isaac F, Crighton KA, Wald P, Exum E, Shurney D, Metz RD, "Do workplace health promotion (wellness) programs work?," Journal of Occupational and Environmental Medicine, 2014 Sep:56(9):927–34. doi: 10.1097/JOM.0000000000000276.

[96] World Health Organization, "MENTAL HEALTH POLICIES AND PROGRAMMES IN THE WORKPLACE," 2005, p. 2, http://www.who.int/mental_health/policy/workplace_policy_programmes.pdf, accessed January 6, 2014.

blood pressure and depression) than engaged employees.[97] The WHO outlined several key workplace stress factors, including:

- "Workload (both excessive or insufficient work);
- Lack of participation and control in the workplace;
- Monotonous or unpleasant tasks;
- Role ambiguity or conflict;
- Lack of recognition at work;
- Inequity;
- Poor interpersonal relationships;
- Poor working conditions;
- Poor leadership and communication;
- Conflicting home and work demands."[98]

Businesses often created or exacerbated employee health problems, especially those associated with stress. For example, customer-facing personnel, such as call center workers, often suffered heavy stress levels from dealing with difficult or unhappy customers.[99] Further, there was increasing evidence that senior executives were often overloaded and needed better time-management and stress-reduction. McKinsey published research in 2016 which suggested that, even at organizations where health was stated as a top priority, executive well-being suffered from high stress levels.[100] Therefore, it was often important for senior executives to speak openly about their own stress challenges and how they managed them.

Businesses endured many negative effects as a result of stress and the mental health challenges among their employees. These included productivity losses, reputational risk, morale risk, and legal costs.[101] Though one study concluded that major depressive disorder (MDD) resulted in 27.2 lost work days per affected employee, the authors found that it was presenteeism, and not absenteeism that had the greatest economic effect.[102] Reputational risk was also a major concern for employers, as the

---

[97] Jim Harter and Amy Adkins, "Engaged Employees Less Likely to Have Health Problems," Gallup, DECEMBER 18, 2015, http://www.gallup.com/poll/187865/engaged-employees-less-likely-health-problems.aspx?utm_source=alert&utm_medium=email&utm_content=morelink&utm_campaign=syndication, accessed May 2016.

[98] World Health Organization, "MENTAL HEALTH POLICIES AND PROGRAMMES IN THE WORKPLACE," 2005, p. 2, http://www.who.int/mental_health/policy/workplace_policy_programmes.pdf, accessed January 6, 2014.

[99] Talkdesk, "8 Stress Management Techniques for Call Center Agents," *talkdesk blog*, https://www.talkdesk.com/blog/8-stress-management-techniques-for-call-center-agents/, accessed May 2016.

[100] Manish Chopra, "Want to be a better leader? Observe more and react less," *McKinsey Quarterly*, February 2016, http://www.mckinsey.com/global-themes/leadership/want-to-be-a-better-leader-observe-more-and-react-less?cid=other-eml-nsl-mip-mck-oth-1603, accessed May 2016.

[101] John A. Quelch and Carin-Isabel Knoop, "Mental Health and the American Workplace" HBS No. 9-515-062 (Boston: Harvard Business School Publishing, 2015).

[102] Howard G. Birnbaum, et al., "Employer Burden of Mild, Moderate, and Severe Major Depressive Disorder: Mental Health Services Utilization and Costs, and Work Performance," Depression and Anxiety 27: 78–89 (2010).

media discussed workplace-related mental health challenges more openly and frequently.[103] Mental health challenges posed a morale risk to the larger employee workforce, as an individual with a mental health issue might act negatively towards or other employees or cause resentment if their work ethic faltered. Finally, employers worried about the legal costs associated with mental health. In the US in 2013, mental disability claims of discrimination based on the Americans with Disabilities Act cost $25.5 million, with the largest number of claims based on depression, anxiety, and manic depressive disorders.[104,105]

Employers responded to these challenges by providing many different types of benefits to their employees. Table 3.1 shows common ways that employees took to help employees manage their stress.

Table 3.1 Steps taken by employers to manage stress

	United States
Promotion of Employee Assistance Programs (EAP)	85 %
Access to financial planning information/services	61 %
Flexible working options	51 %
Expanding EAP services and/or other stress management activities to dependents	46 %
Education and awareness campaigns	40 %
Stress management interventions (e.g., stress management workshops, yoga, tai chi)	39 %
Training for managers	34 %
Specialized training for employees	23 %
External specialist/resources used to design and deliver program(s)	23 %
Risk assessments/stress audits	22 %
Anti-stress space	10 %
Written guidelines on stress	7 %

Source: "The Business Value of a Health Workforce: Staying@Work Survey Report 2013/2014, United States," Towers Watson, January 2014, http://www.towerswatson.com/en/Insights/IC-Types/Survey-Research-Results/2013/12/stayingatwork-survey-report-2013-2014-us, accessed February 2014

[103] Nicholas Kristof, "First Up, Mental Illness. Next Topic Is Up to You," *The New York Times*, http://www.nytimes.com/2014/01/05/opinion/Sunday/kristof-first-up-mental-illness-next-topic-is-up-to-you.html?_r=0, accessed February 2014.

[104] U.S. Equal Employment Opportunity Commission, "ADA Charge Data—Monetary Benefits FY 1997—FY 2013," http://www.eeoc.gov/eeoc/statistics/enforcement/ada-monetary.cfm, accessed February 2014.

[105] John A. Quelch and Carin-Isabel Knoop, "Mental Health and the American Workplace" HBS No. 9-515-062 (Boston: Harvard Business School Publishing, 2015).

Many corporations offered ways to effectively manage stress. For example, Johnson & Johnson offered the Energy for Performance in Life program, which introduced the concept of managing energy, not just time.[106] It encouraged employees to put their energy into purposeful activity, which included areas outside of work.

Other Company-Specific Benefits

While health insurance, workplace wellness programs, and mental health programs were common, corporations often had to provide benefits that were specifically targeted to their employee populations and their unique health challenges. For example, Anglo American, a global mining company, had operations concentrated in Africa, where HIV/AIDs was a significant health challenge for employees and their families. To combat the disease, the company paid for HIV/AIDs testing for their employees and their family members. In 2011, it tested 92 % of employees, up from 10 % in 2003.[107] In addition, the company also covered the cost of treatment for employees and their families and provided HIV/AIDs education to entire villages where the company was located.[108] The company took these actions because it realized that its efforts needed to go beyond testing—for an infectious disease, the company had to impact others in the community. It even provided its suppliers with incentives to test their employees—often migrant workers—in the communities.

Anglo American was not the only company to address the health concerns of workers who were not direct employees. Primark, a division of ABF, launched the HERproject in Bangladesh in 2011.[109] In partnership with NGOs and other global experts, the program focused on providing health care and health education to female workers in garment factories.[110] As of 2016, the program had provided education on hygiene, sexual and reproductive health, maternal health and nutrition to over 5000 female factory workers in Bangladesh.[111]

---

[106] Johnson & Johnson, "Energizing Performance," *JnJ website*, http://www.jnj.com/caring/citizenship-sustainability/our-engagements/energizing-performance, accessed May 2016.

[107] AngloAmerican, "HIV/AIDS Fact Sheet," November, 2012, http://www.angloamerican.com/~/media/Files/A/Anglo-American-Plc/pdf/development/AngloAmerican_HIV_2012.pdf, accessed May 2016.

[108] AngloAmerican, "HIV/AIDS Fact Sheet," November, 2012, http://www.angloamerican.com/~/media/Files/A/Anglo-American-Plc/pdf/development/AngloAmerican_HIV_2012.pdf, accessed May 2016.

[109] Primark, "Programmes: The HERproject," *Primark website*, http://www.primark-bangladesh.com/our-work-in-bangladesh/, accessed May 2016.

[110] Ibid.

[111] Ibid.

After the program, there was an 89 % increase in the number of women who used on-site or local clinics to seek medical advice, 55 % increase in those who used contraception, and a 98 % increase in those who ate animal protein at least once a week.[112] In addition, absenteeism and turnover decreased. After these successes in Bangladesh, Primark expanded the program to China and Southern India with the goal of eventually delivering improved health services and education to all women in its supply chain.[113]

Similarly, Levi Strauss offered its subcontractors training programs for their workers on financial literacy, health and nutrition, and acceptance and equality. Although workers took time off of production lines for the training, the programs increased both productivity and retention.[114] Suppliers who met Levi Strauss's fair labor and environmental practice standards could qualify for low cost loans (subsidized by Levi Strauss) from the International Finance Corporation.[115]

Wage Standards

By the late 2000s, the relationship between wage level and health was well-documented. One group of researchers reviewed a number of studies on this topic and found that people at the lowest socioeconomic status across the world had higher death and illness rates "regardless of whether the major causes of death and disability were from infectious or noninfectious diseases and regardless of how socioeconomic position was measured."[116] Another study found a significant relationship between low wages and increased prevalence of obesity and increased body mass.[117]

By 2015, the minimum wage in the US had become a contentious topic. The effects of raising the minimum wage for the individual worker, companies, and the broader economy had been widely studied, but despite the focus on the topic, they remained fiercely debated. One of the most significant questions was whether rais-

---

[112] Ibid.

[113] Ibid.

[114] Reuters, "The World Bank Group's International Finance Corporation and Levi Strauss & Co. Reward Garment Suppliers for Stronger Labor, Environment, Health and Safety Performance," *Reauters*, November 4, 2014, http://www.reuters.com/article/ca-levi-strauss-ifc-idUSnBw045320a+100+BSW20141104, accessed May 2016.

[115] Ibid.

[116] George Kaplan, Mary Haan, S. Leonard Syme, Meredith Minkler, and Marilyn Winkleby, "Socioeconomic status and health," accessed online at http://www.researchgate.net/profile/George_Kaplan2/publication/30856094_Socioeconomic_status_and_health/links/0fcfd5138d9995a6fb000000.pdf, accessed October 2015.

[117] D. Kim and JP Leigh, "Estimating the effects of wages on obesity," *Journal of Occupational Environmental Medicine*, 2010 May, 52(5):495–500.

ing the minimum wage would lead employers to hire fewer workers, thus increasing unemployment.[118] Researchers with The Center for Economic and Policy conducted a literature review on this topic and found that employment effects were consistently modest.[119] However, research also supported those who argued for the other side. In a brief, researchers with The Cato Institute stated, "a decision to increase the minimum wage is not cost-free; someone has to pay for it, and the research shows that low-skill youth pay for it by losing their jobs, while consumers may also pay for it with higher prices."[120]

Wage decisions were difficult for companies to make, and sometimes had unintended consequences. In February, 2015, Wal-Mart Stores, Inc. ("Walmart"), a multi-national retail corporation, announced that it would raise entry-level wages to at least $9/h in 2015, and to $10/h in early 2016; wages for some manager roles would increase to $13/h and $15/h across the same time period.[121] In October, 2015, Walmart stock fell after announcing at its annual investor meeting that it expected a 6–12 % decline in earnings per share in fiscal year 2017.[122] The disappointing results were attributed to the company's investments in increasing wages and currency exchange rates.[123] Charles Holley, Walmart's executive vice president and chief financial officer stated:

> Fiscal year 2017 will represent our heaviest investment period. Operating income is expected to be impacted by approximately $1.5 billion from the second phase of our previously announced investments in wages and training as well as our commitment to further developing a seamless customer experience…As a result of these investments, we expect earnings per share to decline between 6 and 12 percent in fiscal year 2017, however by fiscal year 2019 we would expect earnings per share to increase by approximately 5 to 10 percent compared to the prior year.[124]

Discussions around wage rates were not limited to the minimum wage. Dan Price, the founder and CEO of Gravity Payments, a Seattle-based credit card

---

[118] John Schmitt, "Why Does the Minimum Wage Have No Discernible Effect on Employment?," *Center for Economic and Policy Research*, February 2013, accessed at http://www.cepr.net/documents/publications/min-wage-2013-02.pdf, accessed October, 2015.

[119] Ibid.

[120] Mark Wilson, "The Negative Effects of Minimum Wage Laws," *Cato Institute*, June 2012, accessed at http://object.cato.org/sites/cato.org/files/pubs/pdf/PA701.pdf, accessed October, 2015.

[121] Doug McMillon, "In Letter to Associates, Walmart CEO Doug McMillon Announces Higher Pay," *Walmart Today* blog, February 15, 2015, Walmart website, http://blog.walmart.com/in-letter-to-associates-walmart-ceo-doug-mcmillon-announces-higher-pay, accessed October 2015.

[122] Lauren Gensler, "Wal-Mart's A Wreck: Stock Suffers Worst Loss In 15 Years On Dim Outlook," *Forbes*, October 14, 2015, http://www.forbes.com/sites/laurengensler/2015/10/14/wal-marts-stock-plunges-on-dim-outlook/, accessed October, 2015.

[123] Ibid.

[124] Walmart, "Walmart strategy drives growth and sustainable returns, Plans $20 billion share repurchase program over two years," *SEC Document October, 2015*, https://www.sec.gov/Archives/edgar/data/104169/000010416915000049/exhibit991-10142015.htm, accessed October 2015.

processing company, made headlines in April 2015 for his decision to increase the salaries of all his employees to a minimum of $70,000.[125] To afford the salary increases, he decided to cut his own salary from nearly $1 million a year to $70,000. His decision was based on research that showed that salary increases, up to around $70,000, could make a large difference in an individual's happiness.[126] Despite the initial positive attention the decision garnered, there were several drawbacks. Though the company gained publicity and additional customers, it also lost customers who found the salary decision political. Although many of the lower-paid workers received salary bumps, the increases were more modest or nonexistent for higher-wage earners, and this factored into two of the company's most valuable employees deciding to quit.[127] Mr. Price said "There's no perfect way to do this and no way to handle complex workplace issues that doesn't have any downsides or trade-offs…I came up with the best solution I could."[128]

Company Structure and Culture

Many of the workplace factors that contributed to employee stress could be attributed to managerial styles and team dynamics. Therefore, a company's hierarchical structure and overarching culture could have far-reaching effects on employee engagement and mental health.

Zappos, an online shoe and retail company, gained media attention in 2015 when it abandoned its traditional, hierarchical management structure and embraced a concept known as "holacracy." Holacracy "removes power from a management hierarchy and distributes it across clear roles, which can then be executed autonomously."[129] At Zappos, employees joined voluntary groups, called "circles," which were comprised of peers who not only helped with difficult projects, but also evaluated and rewarded other employees.[130] When Zappos made the switch, it also offered employees a buyout option for those who were not interested in participating in the new

---

[125] Charles Rile and Poppy Harlow, "Gravity Payments CEO defends $70,000 minimum salary," *CNN Money*, August 10, 2015, http://money.cnn.com/2015/08/09/news/gravity-payments-dan-price-70k-salary/, accessed October 2015.

[126] Ibid.

[127] Patricia Cohen, "A Company Copes with Backlash Against the Raise that Roared," *The New York Times*, July 31, 2015, http://www.nytimes.com/2015/08/02/business/a-company-copes-with-backlash-against-the-raise-that-roared.html, accessed October, 2015.

[128] Ibid.

[129] Holacracy, "How it Works," *Holacracy.org website*, http://www.holacracy.org/how-it-works/, accessed October 2015.

[130] Yuki Noguchi, "Zappos: A Workplace Where No One And Everyone Is The Boss," *NPR*, July 21, 2015, http://www.npr.org/2015/07/21/421148128/zappos-a-workplace-where-no-one-and-everyone-is-the-boss, accessed October 2015.

model; about 14 % of the company's 1600 employees left the company.[131] CEO Tony Hsieh explained the reasoning for the new model, stating:

> Research shows that every time the size of a city doubles, innovation or productivity per resident increases by 15 percent. But when companies get bigger, innovation or productivity per employee generally goes down. So we're trying to figure out how to structure Zappos more like a city, and less like a bureaucratic corporation. In a city, people and businesses are self-organizing. We're trying to do the same thing by switching from a normal hierarchical structure to a system called Holacracy, which enables employees to act more like entrepreneurs and self-direct their work instead of reporting to a manager who tells them what to do.[132]

Many companies also made cultural efforts to engage women in their workforces. According to the WHO, women made up over 40 % of the global labor force and on average earned two-thirds of the income of men.[133] Only 5 % of Fortune 500 CEOs were women.[134] Men and women often perceived workplace stressors differently. A study in the United Kingdom concluded that "women are at risk of increased depression and anxiety if the management style at their workplace is not inclusive or considerate; and male employees are more at risk if they feel excluded from decision making."[135,136]

Representation of women in management varied widely by industry. For example, only 20 % of executives in the technology industry were women.[137] A 2016 article published by The Vitality Institute stated:

> Of the women who are already working in technology, 27 % report feeling as though they are "stalled" in their job and that they are likely to leave within a year, a phenomenon referred to as "a leaky pipeline". So while getting women into the industry is a first step towards gender equity, creating a work environment that fosters their growth and development will be equally important.[138]

---

[131] Ibid.

[132] Zappos, "Holacracy and Self-Organization," *Zappos company website*, http://www.zapposinsights.com/about/holacracy, accessed October 2015.

[133] World Health Organization, "MENTAL HEALTH POLICIES AND PROGRAMMES IN THE WORKPLACE," 2005, http://www.who.int/mental_health/policy/workplace_policy_programmes.pdf, accessed October 2015.

[134] EY, "Women. Fast Forward." *EY website*, PDF publication, http://www.ey.com/Publication/vwLUAssets/ey-women-fast-forward-thought-leadership/$FILE/ey-women-fast-forward-thought-leadership.pdf, accessed October 2015.

[135] Linda Seymour, Sainsbury Centre for Mental Health, "Common mental health problems at work," June 2010, p. 4, http://www.centreformentalhealth.org.uk/pdfs/BOHRF_common_mental_health_problems_at_work.pdf, accessed January 7, 2014.

[136] John A. Quelch and Carin-Isabel Knoop, "Mental Health and the American Workplace" HBS No. 9-515-062 (Boston: Harvard Business School Publishing, 2015).

[137] Gabriela Seplovich and Sarah Kunkle, "The Gender Divide in Tech: A Leaky Pipeline," *The Vitality Institute*, February 23, 2016, http://thevitalityinstitute.org/gender-divide-tech-leaky-pipeline/, accessed May 2016.

[138] Ibid.

Ernst & Young ("EY"), an international professional services and advisory firm, has focused on accelerating women's progress in the workplace through its "Women. Fast forward" initiative. The concept of the program was based on the finding that it would take 80 more years for gender parity in the workplace.[139] The company has both internal and external initiatives focused on advancing women into leadership roles at a faster pace. As part of the program, EY surveyed 400 company leaders in EMEIA, Asia-Pacific and North America on engaging women in the workplace and produced a report highlighting their findings. The report recommended three strategies for accelerating women in the workforce:

1. Illuminate the path to leadership
2. Speed up company culture change with progressive company policy
3. Build supportive environments and work to eliminate conscious and unconscious bias[140]

Due to the lack of female role models in C-suite leadership roles, the first recommendation was focused on highlighting paths of advancement for women, as women could struggle to visualize themselves in those roles. The second area focused on promoting work/life integration and flexibility for women, who shouldered a greater amount of caretaking and housework than men, and the third advocated for a supportive culture, stemming from the CEO downwards.[141]

Conclusion

The role of employers in employee health had evolved significantly over time. By 2016, some employers had become stewards of health and wellness within their workforces, while others met regulatory standards, but remained less involved overall. This note assessed the drivers behind employer investments in employee health, as well as some of the benefits and drawbacks of employee health initiatives. However, several questions remained about the future of employee health.

[139] EY, "Women. Fast Forward," *EY company website*, Women. Fast Forward. Website, http://www.ey.com/GL/en/Issues/Business-environment/women-fast-forward, accessed October 2015.

[140] EY, "Women. Fast Forward," *EY company website*, PDF publication, http://www.ey.com/Publication/vwLUAssets/ey-women-fast-forward-thought-leadership/$FILE/ey-women-fast-forward-thought-leadership.pdf, accessed October 2015.

[141] Ibid.

Discussion Questions on Employee Health

1. **Beyond legally mandated employee health standards, what responsibility do employers have to provide a healthy environment for all employees**?
   (a) In a largely free market economy, employees are able to leave a work environment if they feel another might be healthier for them. Is it therefore the responsibility of the employer to provide a healthy environment, or should the responsibility sit with an individual to find an environment that supports their own personal health priorities?
   (b) Research on individual choice in health care suggests that individuals do not always make the best health decisions for themselves—even when they are educated on the health or financial risks at stake.[142] Therefore, is some level of paternalism from corporations or governments appropriate?
2. **How should corporations balance the economic and social goals of investing in employee health**?
   (a) Many corporate investments in employee health are aimed at generating both corporate and social returns. Are both private and social returns likely from common investments in employee health?
   (b) If positive social returns are generated more broadly, should governments subsidize investments in employee health?
3. **How should corporations consider the investment time horizon in employee health**?
   (a) Given that the potential benefits of employer investments in employee health may outlast employee tenure at that employer, how should corporations prioritize shorter and longer-term investments in employee health?

[142] David A. Asch, MDKevin G. Volpp, MD, "Use Behavioral Economics to Achieve Wellness Goals," December 1, 2014, *Harvard Business Review*, https://hbr.org/2014/12/use-behavioral-economics-to-achieve-wellness-goals, accessed January 2016.

Chapter 4
Community Health

In 2014, Humana Inc., a health insurance provider, announced an enterprise goal of improving the health of the communities it served by 20 %[1] by 2020.[2] It began the initiative in an effort to make healthcare "easier" for individuals, make a complicated and fragmented healthcare system simpler to navigate, and engage individuals in the management of their own health.[3] In its *2015 Progress Report*, Humana stated, "why do we put such emphasis on 'making it easy'? Because as a society, we have made health hard."[4]

To achieve these goals, Humana focused on changing communities. It stated that its goals centered "around communities because we recognize that health care happens in the physician's office and in the hospital, but health happens in the communities in which we live. The health of individuals is influenced—for better or worse—every day by those around us."[5] Research had repeatedly shown that an

Reprinted with permission of Harvard Business School Publishing
Community Health, HBS No. 9-516-075
This case was prepared by John A. Quelch and Emily C. Boudreau.

[1] Measured by the Healthy Days measure, a survey developed by the Centers for Disease Control and Prevention (CDC), which asked participants to self-rate their physical health, mental health, and activity limitation over the previous 30 days.

[2] Humana Inc., "Humana 2020: 2015 progress report," 2015, *Humana website*, PDF publication, accessed at https://closethegap.humana.com/reports/Humana-2020-Progress-Report.pdf, accessed November 2015.

[3] Ibid.

[4] Ibid.

[5] Ibid.

J.A. Quelch, E.C. Boudreau, *Building a Culture of Health*, SpringerBriefs in Public Health, DOI 10.1007/978-3-319-43723-1_4

individual's health was deeply affected by the community in which he or she lived.[6,7] The structural and societal conditions that shaped an individual's environment (e.g., socioeconomic status, education, access to healthy food) were often referred to as the "social determinants of health."[8] (See Exhibit 4.1 for the social determinants of health by category.)

Exhibit 4.1: Social Determinants of Health

Economic Stability	Neighborhood and Physical Environment	Education	Food	Community and Social Context	Health Care System
Employment Income Expenses Debt Medical bills Support	Housing Transportation Safety Parks Playgrounds Walkability	Literacy Language Early childhood education Vocational training Higher education	Hunger Access to healthy options	Social integration Support systems Community engagement Discrimination	Health coverage Provider availability Provider linguistic and cultural competency Quality of care
Health Outcomes Mortality, Morbidity, Life Expectancy, Health Care Expenditures, Health Status, Functional Limitations					

Source: Harry J. Heiman and Samantha Artiga, "Beyond Health Care: The Role of Social Determinants in Promoting Health and Health Equity," *The Kaiser Family Foundation*, November 4, 2015, accessed at http://kff.org/disparities-policy/issue-brief/beyond-health-care-the-role-of-social-determinants-in-promoting-health-and-health-equity, accessed November 2015.

As part of its initiative, Humana invested in several cities across the US, including San Antonio, Texas. In San Antonio, there were over 500,000 Humana members and beneficiaries out of a total population of 1.4 million.[9,10] The company helped establish the San Antonio Health Advisory Board (SAHAB), a group of local physicians and community leaders, as well as a partnership with the city's largest grocer,

[6] Harry J. Heiman and Samantha Artiga, "Beyond Health Care: The Role of Social Determinants in Promoting Health and Health Equity," *The Kaiser Family Foundation*, November 4, 2015, accessed at http://kff.org/disparities-policy/issue-brief/beyond-health-care-the-role-of-social-determinants-in-promoting-health-and-health-equity, accessed November 2015.

[7] The World Health Organization, Commission on Social Determinants of Health, "closing the gap in a generation: Health equity through action on the social determinants of health," 2008, http://apps.who.int/iris/bitstream/10665/43943/1/9789241563703_eng.pdf, accessed November 2015.

[8] Ibid.

[9] Ibid.

[10] United States Census Bureau, "State and County QuickFacts: San Antonio, Texas," accessed at http://quickfacts.census.gov/qfd/states/48/4865000.html, accessed November 2015.

HEB Grocery Company, to address common barriers to eating well.[11] Other local partners included Methodist Healthcare, the San Antonio Food Bank, the mayor's office, the Bexar County Medical Society and San Antonio's Metropolitan Health District.[12] Humana also collaborated with community organizations and providers to create a Health Information Exchange, which tied together the electronic record systems of different providers, enabling physicians to better identify gaps in care.[13]

Though specific results from the work in San Antonio were not yet available in 2015, Humana was focused on measuring and reporting outcomes from its interventions over time. Lisa Stephens, a Director of Measurement for the "Humana 2020" program stated that "measuring our progress is essential to achieving this bold goal. It is how we quantify our commitment to our associates, our members, and the communities where they live."[14]

It used the Healthy Days measure, a survey developed and used by the Centers for Disease Control and Prevention (CDC) for over two decades, which asked participants four simple questions to self-rate their physical health, mental health, and activity limitation over the previous 30 days.[15] In 2014, Humana established baseline measures in the geographic communities where there were a significant number of health plan members, and planned to repeat measurement annually through 2020 using the same methodology.[16] Progress would be measured in the reduction of unhealthy days from baselines.

For Humana, investing in communities like San Antonio was strategically aimed at improving plan members' and employees' health, and thus decreasing Humana's costs. Plan members and employees were influenced by the habits of others in their communities, and improving community health conditions broadly had the potential to encourage their participation in healthier activities. Further, because Humana had a large market share in San Antonio, it was cost efficient to invest in the larger community, rather than target healthcare improvement efforts at only employees and/or plan members.

Humana, had clear incentives for improving community health—contributing to social progress and decreasing healthcare costs of employees and consumers. Many of these same incentives motivated companies outside of healthcare as well. Improving employee wellness through community health was beneficial to any employer that

---

[11] Humana Inc., "Humana 2020: 2015 progress report," *Humana website*, 2015, PDF publication, accessed at https://closethegap.humana.com/reports/Humana-2020-Progress-Report.pdf, accessed November 2015.

[12] Bruce Broussard, "Humana CEO: We're Going To Make The Communities We Serve 20% Healthier," *Forbes*, April 6, 2015, http://www.forbes.com/sites/matthewherper/2015/04/06/humana-ceo-were-going-to-make-the-communities-we-serve-20-healthier/, accessed November 2015.

[13] Bruce Broussard, "Humana CEO: We're Going To Make The Communities We Serve 20% Healthier," *Forbes*, April 6, 2015, http://www.forbes.com/sites/matthewherper/2015/04/06/humana-ceo-were-going-to-make-the-communities-we-serve-20-healthier/, accessed November 2015.

[14] Humana Inc., "Humana 2020: 2015 progress report," 2015, *Humana website*, PDF publication, accessed at https://closethegap.humana.com/reports/Humana-2020-Progress-Report.pdf, accessed November 2015.

[15] Ibid.

[16] Ibid.

provided employer-sponsored health insurance, and improving consumer wellness was a boon for any company that wanted to increase consumer buying power, as chronically ill consumers typically spent a larger portion of their income on healthcare.[17]

However, not all corporations had opted to make such large investments in community health and the role of corporations in combatting community health challenges remained less definitive. Was this even an appropriate corporate responsibility? Why were investments in community health beneficial for corporations? How did corporations differ in their community health investments? And finally, what were the most effective ways that corporations could improve community health?

The Need for Community Health

Life expectancy at birth, which was often used as an overall quality-of-life metric, varied significantly across the world.[18] For example, in 2015, life expectancy at birth in Japan was more than 84 years, while in Afghanistan it was about 51.[19] Such differences existed not only among countries, but across different areas within countries, causing the WHO to state that there was "no necessary biological reason why there should be a difference in [life expectancy at birth] of 20 years or more between social groups in any given country."[20]

Individual health was influenced by many factors including genetic predisposition, health care access, social circumstances, environmental exposure, and individual behavior. Researchers found that genetic predisposition accounted for only about 30 % of the cause of premature death.[21] Though many assumed that the level of healthcare services available impacted individual health, research suggested that on its own, the level of healthcare services was not an especially strong predictor of health outcomes.[22]

However, by 2015, it *was* widely acknowledged that the social and economic conditions within a community had significant impacts on the health of the individuals

[17] Harriet Komisar, "The Effects of Rising Health Care Costs of Middle-Class Economic Security," *AARP Public Policy Institute Issue #74*, 2013, p. 5, accessed at http://www.aarp.org/content/dam/aarp/research/public_policy_institute/security/2013/impact-of-rising-healthcare-costs-AARP-ppi-sec.pdf, accessed November 2015.

[18] Central Intelligence Agency, "The World Factbook," *CIA website*, accessed at https://www.cia.gov/library/publications/the-world-factbook/rankorder/2102rank.html, accessed November 2015.

[19] Ibid.

[20] The World Health Organization, Commission on Social Determinants of Health, "closing the gap in a generation: Health equity through action on the social determinants of health," 2008, http://apps.who.int/iris/bitstream/10665/43943/1/9789241563703_eng.pdf, p. 26, accessed November 2015.

[21] Steven A. Schroeder, M.D., "We Can Do Better—Improving the Health of the American People," *New England Journal of Medicine 357:1221-122*, September 2007, http://www.nejm.org/doi/full/10.1056/NEJMsa073350, accessed November 2015.

[22] Harry J. Heiman and Samantha Artiga, "Beyond Health Care: The Role of Social Determinants in Promoting Health and Health Equity," *The Kaiser Family Foundation*, November 4, 2015, accessed at http://kff.org/disparities-policy/issue-brief/beyond-health-care-the-role-of-social-determinants-in-promoting-health-and-health-equity, accessed November 2015.

who lived within them.[23] Research showed that urban design and the physical environment was associated with physical activity, such as walking and cycling for transportation.[24] Children were more likely to be overweight or obese if they lived in unsafe neighborhoods, in poor housing, and with no access to sidewalks, parks, and recreation centers.[25] Social networks also affected health and opportunity. Social capital "refers to the collective value of all 'social networks' [who people know] and the inclinations that arise from these networks to do things for each other ['norms of reciprocity']."[26] The Harvard Kennedy School emphasized the importance of social capital for improved health and economic outcomes by stating:

> Communities with higher levels of social capital are likely to have higher educational achievement, better performing governmental institutions, faster economic growth, and less crime and violence. And the people living in these communities are likely to be happier, healthier, and to have a longer life expectancy.[27]

However, research by Robert Putnam in the early 2000s had shown that social capital was decreasing.[28] Social media often challenged cohesion in local communities, as children were spending less time playing outside their homes and more time in virtual communities.[29] What's more, in his 2015 book, *Our Kids: The American Dream in Crisis*, he noted an increasing opportunity divide between richer and poorer children in the US. Richer children typically had greater access to social support and extracurricular activities, like sports, which teach children valuable skills such as strong work habits, self-discipline, and a sense of civic engagement.[30] Further, children from poorer communities were more likely to be overweight or obese than children from wealthier communities.[31] Poorer communities were inherently less healthy.

---

[23] Ibid.

[24] Bruce Saelens, James Sallis, and Lawrence Frank, Environmental Correlates of Walking and Cycling: Findings From the Transportation, Urban Design, and Planning Literatures, *The Society of Behavioral Medicine*, Vol 25 Number 2, 2003, accessed at http://www.researchgate.net/profile/Brian_Saelens/publication/10797069_Environmental_correlates_of_walking_and_cycling_findings_from_the_transportation_urban_design_and_planning_literatures/links/09e4150d3905d3541a000000.pdf, accessed November 2015.

[25] Gopal K. Singh, Mohammad Siahpush and Michael D. Kogan, "Neighborhood Socioeconomic Conditions, Built Environments, And Childhood Obesity," *Health Affairs*, 29, no. 3 (2010): 503–512.

[26] Harvard Kennedy School, "About Social Capital," accessed at http://www.hks.harvard.edu/programs/saguaro/about-social-capital/faqs#definition, accessed December 2015.

[27] Ibid.

[28] Robert Putnam, "Bowling Alone: The Collapse and Revival of American Community," *Simon & Schuster, New York*, 2000.

[29] Damian Carrington, "Three-quarters of UK children spend less time outdoors than prison inmates—survey," *The Guardian*, March 25, 2015, http://www.theguardian.com/environment/2016/mar/25/three-quarters-of-uk-children-spend-less-time-outdoors-than-prison-inmates-survey, accessed May 2016.

[30] Michael Jonas, "Opportunity gap," *CommonWealth Magazine*, October 13, 2015, http://commonwealthmagazine.org/economy/opportunity-gap/, accessed December 2015.

[31] Food Research & Action Center, "Relationship Between Poverty and Obesity," *Food Research & Action Center website*, http://frac.org/initiatives/hunger-and-obesity/are-low-income-people-at-greater-risk-for-overweight-or-obesity/, accessed May 2016.

Despite the clear ties between socioeconomic conditions and health, spending on social services in the United States was low compared to other wealthy countries.[32] Authors of the Harvard Business School case, *The State of U.S. Public Health: Challenges and Trends*, stated, "the United States over the last 50 years has focused most of its health resources on providing medical care for individuals after they fall ill. It has placed far less emphasis on the non-medical determinants of health and the prevention of disease for the lives of its citizens."[33]

A Brief History of Corporations and Community Health

It was clear that communities had profound effects on the individuals who lived within them, but what did this have to do with corporations? Intuitively, community health and corporate success were closely related. Companies depended on communities to provide healthy consumers and employees, while communities depended on companies to provide jobs, stability, and wealth-creating opportunities.[34] However, corporate investments in community health had fluctuated over time and place.

Company towns, which appeared in the US and parts of Europe during the Industrial Revolution, were early examples of corporate involvement in community health. These towns were usually built and operated by a single business owner, and often produced goods that were dependent on raw materials (e.g., mining, lumber milling, textile manufacturing, and food production).[35] The business owner owned most of the property in the town, and almost all of the individuals worked for the same enterprise. While some of these towns exploited the local environment and the factory or mill workers, many others had paternalistic company owners who invested in their communities beyond the manufacturing infrastructure, opening schools and parks and providing training to their workforces.[36]

In the US, many large corporations remained place-based throughout the twentieth century and made significant investments in their local communities, especially those where company employees were a large portion of the local population. In

[32] Rosabeth Moss Kanter, Howard Koh, Pamela Yatsko, "The State of U.S. Public Health: Challenges and Trends," HBS No. 9-316-001 (Boston: Harvard Business School Publishing, 2015).

[33] Ibid.

[34] Michael E. Porter and Mark R. Kramer, "Creating Shared Value: How to reinvest capitalism—and unleash a wave of innovation and growth," *Harvard Business Review January–February 2011*, p. 6.

[35] John S. Garner, "The Company Town: Architecture and Society in the Early Industrial Age," *Oxford University Press, New York, New York 1992*, p. 4.

[36] Ibid.

1969, Thomas Watson Jr., the CEO of IBM at the time, said, "…In communities where IBM facilities are located, we do our utmost to help create an environment in which people want to work and live. We acknowledge our obligation as a business institution to help improve the quality of the society we are part of."[37]

IBM was not alone in this sentiment. Companies like Hershey and Kohler developed lively towns that surrounded their companies and provided large workforces.[38,39] However, in the 1980s and 1990s, companies became more dispersed. Outsourcing, the use of part-time workers, and the ability to work remotely (due to the advent of the Internet) all reduced corporations' associations with specific locations. Some of these company towns survived, while others took on new identities or reduced their dependency on single employers or industries.[40]

Throughout the 2000s, business showed an increasing interest in the promotion of human rights, labor standards, and environmental protection.[41] By 2016, there were several indicators that private industry was once again interested in supporting community health initiatives. In the spring of 2014, the Social Enterprise Initiative at Harvard Business School hosted a forum on "Business for Social Impact," which addressed the global role of business in creating social change. The report from that forum stated: "Today's young leaders and the next generation of high net worth individuals aim to build their businesses and invest their wealth in endeavors to fight hunger, bring clean water, education, and healthcare to the world's poor, and end homelessness."[42]

Why Corporations Pursue Community Health

Companies around the world aided community health efforts; however, not all corporations pursued community health activities for the same reasons. Some companies were driven by a sense of corporate duty and obligation while others were motivated by the desire for cost reduction (e.g., mitigating employee health costs through community health investments). By 2015, several organizations and

[37] IBM Archives, "Valuable resources on IBM's history," accessed at http://www-03.ibm.com/ibm/history/ then selected the PDF entitled "Quintessential quotes" which can be accessed at http://www-03.ibm.com/ibm/history/documents/pdf/quotes.pdf, accessed November 2015.

[38] Tanya Mohn, "The Evolution Of Company Towns: From Hershey's To Facebook," *Forbes*, January 17, 2013, http://www.forbes.com/sites/tanyamohn/2013/01/17/the-evolution-of-company-towns-from-hersheys-to-facebook/, accessed November 2015.

[39] Kohler Village, "Village History," *Kohler Village town website*, http://kohlervillage.org/visitors/village-history, accessed November 2015.

[40] Tanya Mohn, "The Evolution Of Company Towns: From Hershey's To Facebook," *Forbes*, January 17, 2013, http://www.forbes.com/sites/tanyamohn/2013/01/17/the-evolution-of-company-towns-from-hersheys-to-facebook/, accessed November 2015.

[41] Ibid.

[42] Harvard Business School, "Business For Social Impact: Forum Summary," accessed at http://www.hbs.edu/socialenterprise/Documents/BUSIForum2014Summary.pdf, accessed November 2015.

researchers had catalogued the reasons why businesses invested in community health.[43,44] (See Exhibit 4.2 for the CDC's list of reasons why businesses invested in community health.)

Exhibit 4.2: CDC: Business Case for Investments in Community Health

- Improve the health status, and therefore the productivity, of an employer's current and future workforce.
- Control direct (health care) and indirect (absenteeism, disability, presenteeism) costs to the employer.
- Create both the image and the reality of a healthy community that may help recruitment and retention of workforce talent in tight labor markets.
- Increase the buying power and consumption level for business products, in particular nonmedical goods and services, by improving the health and wealth of a community.
- Strengthen an employer's brand and recognition in the community.
- Generate, for individual business leaders, positive feelings of civic pride and responsibility and of being a constructive member of the community.
- Channel corporate philanthropy in a direction that will improve community relations, goodwill, or branding with the potential for a positive return for the business enterprise itself.
- Help create public and private partnerships and a multistakeholder community leadership team that can become the foundation for collaboration, cooperation, and community-based problem solving for many other issues affecting the business community, such as economic development and education.

Source: Andrew Webber and Suzanne Mercure, "Improving population health: the business community imperative," Prev Chronic Dis 2010;7(6):A121, accessed at http://www.cdc.gov/pcd/issues/2010/nov/10_0086.htm, accessed November 2015.

In general, this literature could be summarized by four main drivers:

Complying with regulation: To achieve regulatory compliance, both in relevant public health arenas as well as more broadly, many companies hired Chief Compliance Officers and other compliance staff. A PwC compliance survey in 2015 showed that corporate compliance staffing and budgets were trending up.[45] (See Exhibit 4.3 for the growth in compliance staffing and budgeting across industries.)

---

[43] Andrew Webber and Suzanne Mercure, "Improving population health: the business community imperative," Prev Chronic Dis 2010;7(6):A121, accessed at http://www.cdc.gov/pcd/issues/2010/nov/10_0086.htm, accessed November 2015.

[44] Nicolaas P. Pronk, PhD, Catherine Baase, MD, Jerry Noyce, MBA, and Denise E. Stevens, PhD, "Corporate America and Community Health: Exploring the Business Case for Investment," J Occup Environ Med. 2015 May;57(5):493–500.

[45] PWC, "PWC State of Compliance Survey 2015: Staffing and budgets are rising," accessed at http://www.pwc.com/us/en/risk-management/state-of-compliance-survey/staffing-budgets-rising.html, accessed November 2015.

Exhibit 4.3: Growth in Compliance Staffing and Budgeting Across Industries

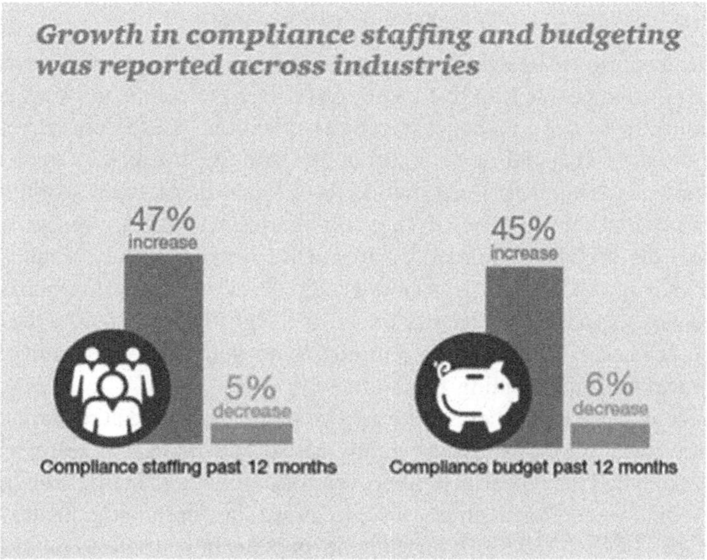

Growth in compliance staffing and budgeting was reported across industries

47% increase

5% decrease

Compliance staffing past 12 months

45% increase

6% decrease

Compliance budget past 12 months

Source: PWC, "PWC State of Compliance Survey 2015: Staffing and budgets are rising," accessed at http://www.pwc.com/us/en/risk-management/state-of-compliance-survey/staffing-budgets-rising.html, accessed November 2015.

Improving social responsibility: Corporate support of community health was often also part of broader efforts to become more socially responsible. Corporations pursued social responsibility for a number of reasons, including out of moral obligation. Investments also provided businesses with brand differentiation, consumer engagement, and employee engagement opportunities.[46] Additionally, socially responsible investments in community health often created beneficial ties between corporations and other organizations in the community.[47]

Amending transgressions: For many corporations, community health activities were a way of amending past transgressions and rebuilding trust within a community. In a 2006 *Harvard Business Review* article, Michael Porter also pointed out that "heightened corporate attention to [corporate social responsibility] has not been entirely voluntary. Many companies awoke to it only after being surprised by public responses to issues they had not previously thought were part of their business

[46] James Epstein-Reeves, "Six Reasons Companies Should Embrace CSR," *Forbes*, February 21, 2012, accessed at http://www.forbes.com/sites/csr/2012/02/21/six-reasons-companies-should-embrace-csr/, accessed November 2015.

[47] Andrew Webber and Suzanne Mercure, "Improving population health: the business community imperative," Prev Chronic Dis 2010;7(6):A121, accessed at http://www.cdc.gov/pcd/issues/2010/nov/10_0086.htm, accessed November 2015.

responsibilities."[48] In other words, companies became more socially responsible to mitigate the publicity risks of not being so, especially as activist organizations grew more effective at raising public pressure against corporations.[49]

Promoting employee health: A corporate investment in a community was really an investment in the people who lived within it—often the company's own employees. The Vitality Institute, a global research organization focused on the reduction of chronic disease risk, published a report in 2015 on the linkages between workforce and community health. It stated that "efforts focused solely on workplace health promotion and/or increased health care cost sharing with employees are insufficient to address the broader, community-based drivers that influence employees' individual behaviors outside of the workplace."[50] Even when healthy behaviors were modeled and supported in the workplace for eight hours per day, these benefits could be negated if the employee returned to an unhealthy community and home environment for the remaining 16 h of the day.

Therefore, these investments were also strategically aimed at reducing employee healthcare costs, which had risen steadily across the 2000s.[51] Investments in community health had the potential to improve productivity and increase engagement of current employees. Corporations also invested in community health to ensure longer-term stability. Upstream investments in education produced more skilled and knowledgeable workers in the future.[52] For example, IBM invested in the training and recruitment of underserved youths through its Pathways in Technology Early College High Schools. [53]

Enhancing consumer health: Businesses were not only dependent on healthy employees to produce goods and services, they were also dependent on consumers to use what they produced. Therefore, corporations invested in community health to promote consumer health. Chronically ill individuals and families often had less

[48] Michael E. Porter and Mark R. Kramer, "Strategy and Society: The Link Between Competitive Advantage and Corporate Social Responsibility," *Harvard Business Review December 2006*, p. 2.

[49] Ibid.

[50] Vera Oziransky, Derek Yach, Tsu-Yu Tsao, Alexandra Luterek, Denise Stevens, "Beyond the Four Walls: Why Community Is Critical to Workforce Health," *The Vitality Institute*, July 2015, accessed at http://thevitalityinstitute.org/site/wp-content/uploads/2015/07/VitalityInstitute-BeyondTheFourWalls-Report-28July2015.pdf, accessed November, 2015.

[51] The Kaiser Family Foundation and Health Research & Educational Trust, "Employer Health Benefits: 2015 Summary Findings," September 22, 2015, http://kff.org/health-costs/report/2015-employer-health-benefits-survey/, accessed October 2015.

[52] Rosabeth Moss Kanter, "From Spare Change to Real Change: The Social Sector as Beta Site for Business Innovation," Harvard Business Review May–June 1999, p. 124.

[53] Alexandra Luterek and Vera Oziransky, "IBM: Redesigning the High School to Workforce Pipeline," *The Vitality Institute*, http://thevitalityinstitute.org/ibm-redesigning-the-high-school-to-workforce-pipeline/, accessed May 2016.

discretionary income, as these consumers used a larger portion of their income on healthcare expenses.[54] Keeping community members well increased their purchasing power. One study on the financial burden of significant medical care stated:

> A significant illness or injury can lead to sizable medical bills for people who are uninsured and for people who have health insurance but incur high deductible or other uncovered expenses. In the first half of 2011, one-third of people (of all ages) were in a family experiencing a financial burden from medical care, including families that had problems paying medical bills (20 percent) or are currently paying bills over time. One in ten people are in a family with medical bills it cannot pay at all.[55]

Table 4.1 How community health organizations benefit from corporate partnership

Benefit	Description
Manage budget constraints	Non-profit community organizations typically had smaller budgets, and large corporations often had philanthropic foundations that could scale their efforts
Build capacity	Businesses offered skills-based volunteering to their employees and other pro bono services to help non-profit community organizations innovate (e.g. in marketing, design, supply chain)
Leverage technological capacity	Partnering with technology companies offered community health groups the ability to more effectively scale and measure health interventions
Reduce the potential negative impact of business practices	Community organizations could ask for transparency and accountability from private corporations they worked with on health programs (e.g., learning about new products that could affect health)

Source: Vera Oziransky, Derek Yach, Tsu-Yu Tsao, Alexandra Luterek, Denise Stevens, "Beyond the Four Walls: Why Community Is Critical to Workforce Health," The Vitality Institute, July 2015, accessed at http://thevitalityinstitute.org/site/wp-content/uploads/2015/07/VitalityInstitute-BeyondTheFourWalls-Report-28July2015.pdf, accessed November, 2015

Why Communities Seek Corporate Partnership

There were also many reasons why communities sought out corporate partners for health-promoting efforts. Community organizations had different challenges and goals, and therefore looked for diverse capabilities in their corporate partners. Table 4.1 shows the Vitality Institute's findings as to how community health organizations benefit from partnerships with private corporations.

---

[54] Harriet Komisar, "The Effects of Rising Health Care Costs of Middle-Class Economic Security," *AARP Public Policy Institute Issue #74*, 2013, p. 5, accessed at http://www.aarp.org/content/dam/aarp/research/public_policy_institute/security/2013/impact-of-rising-healthcare-costs-AARP-ppi-sec.pdf, accessed November 2015.

[55] Harriet Komisar, "The Effects of Rising Health Care Costs of Middle-Class Economic Security," *AARP Public Policy Institute Issue #74*, 2013, p. 5, accessed at http://www.aarp.org/content/dam/aarp/research/public_policy_institute/security/2013/impact-of-rising-healthcare-costs-AARP-ppi-sec.pdf, accessed November 2015.

How Corporations Impact Community Health

Ways Corporations Pursue Community Health

Corporations engaged with communities in different ways depending on factors such as their capabilities, size, location, employee-base, and consumers. Furthermore, not all corporations affected the same types of communities. Some communities were groups of people who lived in the same geographic area; however, communities could also form as a result of common interests, personal traits, or other defining factors.

For example, the advent of the internet propagated virtual communities, where like-minded individuals could connect online. PatientsLikeMe was an online community where patients could find others with the same disease, share their experiences, and learn from one another.[56] In its first five years, the site grew to include 15 different patient communities where over 80,000 patients discussed 19 diseases.[57] The site was intended to provide social support to its users, similar to more traditional, place-based communities.[58]

The following encompassed the wide range of corporate activities in community health:

Providing basic community services: In some communities, governments provided most basic health and social services—making corporate involvement less necessary. However, in others, these services were lacking. In emerging economies, where government-funded services were often less robust, many corporations invested in basic community health services to promote development. These kinds of investments often buoyed the corporation and the community, as health and social services were prerequisites for larger-scale socioeconomic development and public health barriers could make business operations more difficult or deplete a local workforce.

BASF India Limited, a division of BASF, a multinational chemical company, took this approach and focused its CSR activities in three main service areas: (1) Community drinking water supply, (2) Sanitation facilities (with focus on toilets) and (3) Education.[59] BASF India stated that in all of these areas, "the Company will strive to create shared value for BASF and the communities in which it operates."[60] In March, 2015, it opened a community drinking water plant in Singaperumal Koil (SP Koil), a village near a BASF plant in Chennai, and reached around 1500 individuals in the first 3 weeks.[61]

[56] Sunil Gupta and Jason Riis, "PatientsLikeMe: An Online Community of Patients," HBS No. 9-511-093 (Boston: Harvard Business School Publishing, 2012).

[57] Ibid.

[58] Ibid.

[59] BASF India Limited, "CSR Policy," accessed at https://www.basf.com/documents/in/en/investor-relations/code-of-conduct-and-other-policies/CSR_Policy.pdf, accessed November 2015.

[60] Ibid.

[61] https://www.basf.com/in/en/company/sustainability/corporate-social-responsibility/community-drinking-water-plant.html

Creating or funding community health programs: One of the most common ways that corporations pursued community health was through the provision of health programs. In communities where public entities covered most basic health and social services, corporations often created community health programs that supplemented basic services. These included, among other things, programs aimed at improving education, care for specific diseases, or increasing public health awareness.

Univision Communications Inc. ("Univision"), the leading media company serving Hispanic Americans, launched "Univision Contigo" (Univision With You).[62] Through the program, Univision increased awareness of key issues among the Hispanic American community and provided resources to help Latino families take action. The program was well-aligned with their main area of business, and showcased how community health investments were not always geographically-based. Univision focused on health as part of the "building blocks of success"—its Univision Contigo website stating, "these resources will fall under the four universally accepted building blocks of success: Education, Health, Prosperity and Participation."[63]

In 2013, it educated Hispanic Americans on the Affordable Care Act (ACA) and encouraged them to enroll in health insurance.[64] At the time, US Census data had shown that about one in three Hispanics lacked health coverage and many had low to moderate incomes that could qualify them for Medicaid or federal subsidies; however, most Hispanics were unfamiliar with the health benefits under the ACA.[65] In 2015, Univision Contigo also offered healthy menu planning as part of its continuing health promotion efforts.[66]

Oftentimes, corporations that had strong ties to certain geographic areas directed their community health programs more locally. HEB Grocery Company ("HEB"), the largest food retailer and one of the largest privately held companies in Texas, with nearly 80,000 employees, more than 300 stores, and over 50 % market share in many cities and towns.[67] Given the significant role that food played in health and the

[62] Univision Corporate, "Univision Communications Inc. Launches "Univision Contigo" Expanding Outreach & Investment to Empower Hispanic America," *Univision Press Release*, November 25, 2013, accessed at http://corporate.univision.com/2013/11/univision-communications-inc-launches-univision-contigo-expanding-outreach-investment-to-empower-hispanic-america/, accessed November 2015.

[63] Univision Contigo, "About," *Univision Contigo website*, accessed at http://www.univisioncontigo.com/en/about/, accessed November 2015.

[64] David Morgan, "WellPoint, Univision team up to explain Obamacare to Hispanics," *Reuters*, June 27, 2013, accessed at http://www.reuters.com/article/2013/06/27/us-usa-healthcare-hispanics-idUSBRE95Q19X20130627#MghfaheDt3la09VE.97, accessed November 2015.

[65] Ibid.

[66] Univision Contigo, *Program website*, accessed at http://www.univisioncontigo.com/en/, accessed November 2015.

[67] Sean Lester, "What is H-E-B? The history behind the Butt family and how the stores became a Texan grocery staple, *The Dallas Morning News*, March 25, 2014, http://bizbeatblog.dallasnews.com/2015/03/what-is-h-e-b-the-history-behind-the-butt-family-and-how-the-stores-became-a-texan-grocery-staple.html/, accessed May 2016.

health challenges within Texas (e.g., more than two-thirds of the population was diabetic or pre-diabetic), the grocer realized it had a unique role to play in the community.[68]

HEB considered three main ways to build community health: first, focus on employees; second on consumers; and third on the broader community. For its employees, it improved its health insurance offerings, offered education services, and encouraged the use of primary healthcare services.[69] To empower consumers to make healthier decisions, the company conducted a product ingredient analyses of its food products and made efforts to improve food labeling. The company used its in-store pharmacies to extend the continuum of care—helping to lower hospital admission rates, screening for free about 250,000 people a year for certain health conditions in-store, and offering in-store nutrition counseling (e.g., education for newly diagnosed diabetics).[70] Its programs were expansive, but given its market coverage and brand visibility, it made sense for HEB to invest in community health. A company with a smaller market share might have more difficulty justifying such investments.

Mitigating community health crises: Corporate influence in community health was not always planned. Sometimes, natural or health disasters made corporate influence all but necessary for continuous business operations or employee safety. In 2014, ArcelorMittal, the largest steel company in the world, stumbled into a community health epidemic—an outbreak of Ebola, a deadly virus. The company realized that the outbreak threatened the economic and social situation in Liberia, where it operated an iron-ore mine.[71]

The government's ability to contain the situation was wearing thin, and the company understood that it could not contain the outbreak by focusing only on its own employees; it had to take an active role in containing the situation by working with the community.[72] ArcelorMittal contacted local NGOs, worked with the government, and donated hospital supplies and equipment (e.g., gloves, chlorine, and personal protective equipment) to contain the virus and treat patients.[73] It also hired a South African Ebola expert, who spent a month educating ArcelorMittal employees, community members, and healthcare workers.[74]

---

[68] HEB, "LIFE: Live Well," *HEB website*, https://www.heb.com/static-page/article-template/LIFE-Live-Well, accessed May 2016.

[69] HEB, "Partner Services," *HEB website*, https://www.heb.com/static-page/H-E-B-partner-services, accessed May 2016.

[70] HEB, "LIFE: Live Well," *HEB website*, https://www.heb.com/static-page/article-template/LIFE-Live-Well, accessed May 2016.

[71] Erika Fry, "Business in the hot zone: How one global corporation has managed the Ebola epidemic," *FORTUNE*, October 30, 2014, http://fortune.com/2014/10/30/arcelormittal-business-liberia-ebola-outbreak/, accessed November 2015.

[72] Ibid.

[73] Ibid.

[74] Ibid.

Creating new business models: Other companies even built community health initiatives into their business models. Warby Parker, a prescription eyeglass and sunglass company, sold designer eyewear and ensured that for every pair of glasses sold, a pair was distributed to one of the one billion people worldwide who lacked access to glasses.[75] This "one-for-one business model," where companies promised to distribute a product free of charge for each product sold, was utilized by many companies starting in the mid-2000s.[76] While the model assured significant social impact, critics noted that it could undermine local businesses in developing countries, where individuals often received the goods, and create poor self-image among the recipients.[77]

To address this controversy, Warby Parker opted to donate to nonprofit partners that trained men and women in developing countries to sell the glasses within their local communities. Warby Parker explained their policy with the statement: "Donating is often a temporary solution, not a lasting one…Instead of donating, our partners train men and women to sell glasses for ultra-affordable prices, which allows them to earn a living. More important, it forces our partners to offer glasses that people actually want to buy…"[78]

Re-purposing business strengths: Other corporations used their traditional business acumen in new ways to benefit community health. In 2009, Intel Corporation ("Intel") had rising employee healthcare costs, even after it introduced consumer-driven healthcare insurance plans and employee wellness programs.[79] These initiatives resulted in increased employee engagement and awareness; however, they were not enough to stem the increasing costs.

Intel formulated a creative solution—it created a pilot program, the Healthcare Marketplace Collaborative (HMC), in Portland, Oregon and partnered with local healthcare providers to redesign how care was provided for all patients at the participating hospitals.[80] Intel used its strategic expertise in supply chain management to improve treatment paradigms for conditions such as diabetes and lower back pain. Results showed that treatment costs for three conditions fell by 29–49 % and patient satisfaction improved.[81] The results were positive for Intel and created social progress for all citizens in Portland.

---

[75] Warby Parker, "History," *Warby Parker website*, accessed at https://www.warbyparker.com/history, accessed November 2015.

[76] Knowledge@Wharton, "The One-for-one Business Model: Avoiding Unintended Consequences," *Wharton, University of Pennsylvania*, February 16, 2015, http://knowledge.wharton.upenn.edu/article/one-one-business-model-social-impact-avoiding-unintended-consequences/, accessed November 2015.

[77] Ibid.

[78] Warby Parker, "Buy-a-Pair-Give-a-Pair," *Warby Parker website*, accessed at https://www.warbyparker.com/buy-a-pair-give-a-pair, accessed November 2015.

[79] Patricia A. McDonald, Robert S. Mecklenburg, and Lindsay A. Martin, "The Employer-Led Health Care Revolution," *Harvard Business Review July–August 2015.*

[80] Ibid.

[81] Ibid.

Similarly, IBM, also a technology company, offered scientists and other researchers use of the World Community Grid, one of the largest virtual supercomputers. IBM designed the World Community Grid to pool the unused computing power of millions of personal computers worldwide, and then made it available to scientists for timely research.[82] It was designed to aid research efforts focused on difficult humanitarian projects (e.g., new treatments for HIV/AIDS, cancer research, and affordable water purification).[83]

Business strengths included the expertise and skills of a company's employees. Many corporations supported communities by valuing and encouraging employee-involvement in community health. While most corporations had volunteer programs, some companies took these efforts further. At IBM, more than 3000 employees had worked in the developing world through the company's Corporate Service Corps program; these donations amounted to $50 M worth of employee time by 2016.[84] The program created value for the communities it served and boosted community support of IBM. In addition, retention rates of participating employees were higher than those of non-participants.[85]

Investing for community health: Impact investing was an innovation that sought to create both social and economic returns. On impacting investing, the summary report from the Harvard Business School "Business for Social Impact Forum" conference stated:

> An enormous amount of capital exists in the world, and much of it is trapped—earning low returns and providing no social impact. With impact investing, there is the opportunity to unlock and reallocate massive amounts of capital, providing social entrepreneurs the capital they need to scale successful social innovations that can transform entire sectors, while simultaneously delivering investors with attractive and uncorrelated returns. Impact investing requires financial innovations (such as social impact bonds), rules (such as an impact accounting system), measures of social outcomes, and changing the mindset of investors and philanthropists.[86]

Though some organizations sought to use these investments to drive improvement in community health and create returns for their corporations, their use remained limited for several reasons. Social impact bonds (SIBs), a specific kind of impact investment, were intended to provide high-quality services to at-risk com-

---

[82] IBM: News Releases, "In Watson's Wake, IBM World Community Grid Registration Skyrockets 700 %," *IBM website*, February 18, 2011, https://www-03.ibm.com/press/us/en/pressrelease/33752.wss, accessed May 2016.

[83] Ibid.

[84] PYXERA Global, "IBM Corporate Service Corps," https://www.pyxeraglobal.org/ibm-corporate-service-corps/, accessed May 2016.

[85] Gina Tesla, "IBM Uses the Corporate Service Corps to Attract, Develop and Retain Talent," *The New Global Citizen*, May 2, 2016, https://www.newglobalcitizen.com/global-pro-bono/6654, accessed, May 2016.

[86] Harvard Business School, "Business For Social Impact: Forum Summary," accessed at http://www.hbs.edu/socialenterprise/Documents/BUSIForum2014Summary.pdf, accessed November 2015.

munities more efficiently than more traditional, government programs.[87] Then, a portion of the savings from improved efficiency would be returned to the private investors as interest. However, SIBs contrasted with traditional bonds in that investors risked losing their initial investment if the SIB did not have the expected social results.[88]

While the increased level of risk deterred some corporations from using SIBs, other cases suggested that outcomes measurement for impact investments also present barriers to appropriate use. In 2015, Goldman Sachs announced that its investment in a Utah preschool program had helped over one hundred kindergarten students avoid special education.[89] It received a $260,000 payment in return, and was expected to receive additional payouts over time.[90] Though the Governor of Utah, Gary R. Herbert was positive about the program, The New York Times reported that education experts had questioned the measurement of outcomes in the program, stating:

> Nine early-education experts who reviewed the program for The New York Times quickly identified a number of irregularities in how the program's success was measured, which seem to have led Goldman and the state to significantly overstate the effect that the investment had achieved in helping young children avoid special education.[91]

Avoiding unintended consequences: Corporations could not predict all of the effects of a business decision, and there were often unintended public health consequences. This phenomenon was not limited to businesses and the local communities within which they operated. As economies globalized, corporations were forced to consider the unintended community health consequences that could result from worldwide supply chains.

This was evident in corporate sourcing decisions in the palm oil industry. In 2015, Indonesia and Malaysia produced about 85 % of the global supply of palm oil, a vegetable oil derived from the fruit of oil palm, and hired around 3.5 million workers to maintain the palm oil plantations.[92] Palm oil was a $30 billion global indus-

---

[87] John A. Quelch and Margaret L. Rodriguez, "Fresno's Social Impact Bond for Asthma," HBS No. 9-515-028 (Boston: Harvard Business School Publishing, 2014).

[88] John A. Quelch and Margaret L. Rodriguez, "Fresno's Social Impact Bond for Asthma," HBS No. 9-515-028 (Boston: Harvard Business School Publishing, 2014).

[89] Nathaniel Popper, "Success Metrics Questioned in School Program Funded by Goldman," *The New York Times*, November 3, 2015, http://www.nytimes.com/2015/11/04/business/dealbook/did-goldman-make-the-grade.html?src=me&_r=0, accessed November 2015.

[90] Nathaniel Popper, "Success Metrics Questioned in School Program Funded by Goldman," *The New York Times*, November 3, 2015, http://www.nytimes.com/2015/11/04/business/dealbook/did-goldman-make-the-grade.html?src=me&_r=0, accessed November 2015.

[91] Ibid.

[92] Laura Villadiego, "Palm oil: why do we care more about orangutans than migrant workers?," *The Guardian*, November 9, 2015, http://www.theguardian.com/sustainable-business/2015/nov/09/palm-oil-migrant-workers-orangutans-malaysia-labour-rights-exploitation-environmental-impacts, accessed November 2015.

try.[93] The industry had been criticized for its effects on the environment and its exploitation of the palm oil plantation workers.[94]

In Malaysia, many of the plantation laborers were migrants from the Philippines, Nepal, Bangladesh and Indonesia, and many complained of harsh working conditions that left them indebted and overworked, with many dying from exhaustion, disease, or beatings.[95] Eric Gottwald, the legal and policy director at the International Labor Rights Forum said, "It is a very abusive system that includes labour-trafficking, debt bondage and unfair payments. A lot of those workers are undocumented and Malaysian law is very unfriendly to migrants."[96]

An industry group called the Roundtable on Sustainable Palm Oil (RSPO) was created to certify the practices of palm oil companies, though a producer's membership in the RSPO was voluntary.[97] Many companies that used palm oil in their products took part in the RSPO and also took additional action on their own to ensure sustainable environmental and fair labor practices. Unilever, a multinational consumer goods company, was a founding member of RSPO and made a commitment to trace all of the palm oil it used in its European Foods business to sustainable plantations, reaching that goal at the end of March, 2015.[98]

Barriers to Corporate Involvement in Community Health and Potential Solutions

Though the reasons motivating corporations to invest in community health were significant, there were also several barriers that business leaders and researchers had identified. In 2014, HERO convened more than 50 executives at the HealthPartners Institute for Education and Research, a non-profit dedicated to improving health, to conduct an assessment of employer leadership in community health. During the

[93] SYED ZAIN AL-MAHMOOD, "Palm-Oil Migrant Workers Tell of Abuses on Malaysian Plantations," *The Wall Street Journal*, July 26, 2015, http://www.wsj.com/articles/palm-oil-migrant-workers-tell-of-abuses-on-malaysian-plantations-1437933321, accessed November 2015.

[94] Anthony Kuhn, "Palm Oil Plantations Are Blamed For Many Evils. But Change Is Coming," *NPR*, April 21, 2015, http://www.npr.org/sections/goatsandsoda/2015/04/21/396815303/palm-oil-plantations-are-blamed-for-many-evils-but-change-is-coming, accessed November 2015.

[95] SYED ZAIN AL-MAHMOOD, "Palm-Oil Migrant Workers Tell of Abuses on Malaysian Plantations," *The Wall Street Journal*, July 26, 2015, http://www.wsj.com/articles/palm-oil-migrant-workers-tell-of-abuses-on-malaysian-plantations-1437933321, accessed November 2015.

[96] Laura Villadiego, "Palm oil: why do we care more about orangutans than migrant workers?," *The Guardian*, November 9, 2015, http://www.theguardian.com/sustainable-business/2015/nov/09/palm-oil-migrant-workers-orangutans-malaysia-labour-rights-exploitation-environmental-impacts, accessed November 2015.

[97] Anthony Kuhn, "Palm Oil Plantations Are Blamed For Many Evils. But Change Is Coming," *NPR*, April 21, 2015, http://www.npr.org/sections/goatsandsoda/2015/04/21/396815303/palm-oil-plantations-are-blamed-for-many-evils-but-change-is-coming, accessed November 2015.

[98] Unilever, "Transforming the palm oil industry," *Unilever website*, https://www.unilever.com/sustainable-living/what-matters-to-you/transforming-the-palm-oil-industry.html, accessed November 2015.

Table 4.2 Most important barriers and limitations that may keep employers from playing a critical role in improving community health

Barrier	Description
Lack of understanding	This includes a lack of understanding of reasons to care about health outside of the business' four walls, of what "health" actually represents, of the diverse agendas of stakeholders involved and their potential misalignment, of ideology, of who is responsible, of the benefit, of what is actually being asked for
Lack of strategy or "playbook"	There is no framework or model that speaks to business needs. There is no "playbook" that outlines what business should do and how it should be done. The lack of a common language and definitions was noted here as well. Not knowing where to start, what kind of infrastructure to create, and how to get other businesses on board. Not knowing how to convene the community and other stakeholders
Complexity of the problem	The vision of what is needed is so large that it almost feels "not doable." The expressed need to be able to "walk before you run" and build internal worksite health capacity and capability first (invest in your own employees) and then go into the community. The problem is so large that it needs to be made simpler so that the scope and complexity are not associated with long time frames and high risk of failure. Complexity of the collaboration needs to be handled by conveners with high levels of expertise
Trust	Companies may not be willing to take the risk of being a first-mover. Lack of a trusted convener who can bring many stakeholders with varying interests together and facilitate ongoing progress in the initiative
Lack of resources, time, and leadership	Small businesses may not have the time or resources to invest like other, larger organizations may do. On the contrary, small business may already be closely connected to their local communities by way of how they operate. Lack of a sense of urgency, which delays decision making. A large upfront investment of time, resources, and money with a potential payoff lagging for years may not sway business leaders to act
Policies and regulations	Alignment with federal, state, and local policies and regulations needs to be checked, and gap analyses need to point out where changes are needed. Policies and regulations that provide corporate incentives to provide leadership and resources for community health are needed
Leadership philosophy	Lack of a unifying leadership philosophy that can span various views on how employee and community health may be linked

Source: Nicolaas P. Pronk, PhD, Catherine Baase, MD, Jerry Noyce, MBA, and Denise E. Stevens, PhD, "Corporate America and Community Health: Exploring the Business Case for Investment," J Occup Environ Med. 2015 May;57(5):493–500

session, participants identified a number of barriers to corporate-led investments in community health.[99] Table 4.2 shows the barriers the group identified.

The following four corporate activities were helpful to overcome the identified barriers: (1) alignment, (2) partnership with community organizations, (3) coordination among corporations, and (4) measurement protocols.

[99] Nicolaas P. Pronk, PhD, Catherine Baase, MD, Jerry Noyce, MBA, and Denise E. Stevens, PhD, "Corporate America and Community Health: Exploring the Business Case for Investment," J Occup Environ Med. 2015 May;57(5):493–500.

Alignment: Alignment referred to the idea of finding commonalities between corporate investments in community health and a company's main area of business. Alignment had not always been considered a benefit. Historically, philanthropic endeavors, such as corporate gifts to community health organizations, were considered distinct from a company's main area of business, and throughout the 1950s and 1960s, many large corporations began to give to social causes through separate corporate foundations.[100] At this time, corporations rarely concentrated their philanthropic efforts in one area of giving, and they often gave to causes completely unrelated to their core business.[101]

However, over time, corporations saw the benefits of aligning their social endeavors with their main area of business.[102] In 2015, a *Harvard Business Review* article argued that "there is an increasing pressure to dress up [corporate social responsibility] as a business discipline and demand that every initiative deliver results. That is asking too much of [corporate social responsibility] and distracts from what must be its main goal: to align a company's social and environmental activities with its business purpose and values."[103] In other words, a corporation's actions to advance social causes need not always require a return on investment for the corporation, but they should align with a corporation's main area of business and mission.

Alignment was simpler for some companies, such as those in the healthcare sector, than for others, say those in the technology sector. However, there were many companies outside of the healthcare industry that used their business strengths to advance community health causes. After a devastating earthquake in Nepal in 2015, FedEx Corporation ("FedEx"), a global transportation and logistics company, committed approximately $1 million in cash, transportation support, and a chartered flight service to deliver medical aid and supplies to victims of the natural disaster.[104] Dr. Karen Reddington, president, FedEx Express, Asia Pacific said, "As a global express logistics provider, we're in a great position to help in the relief effort by leveraging our network and our relationships with a number of important NGOs which are already working to help Nepal cope with the quake and its aftermath."[105] The company had existing relationships with Direct Relief, Heart to Heart International, Water Missions International, American Red Cross and The Salvation Army, which it utilized quickly and efficiently to reach those who were affected.

---

[100] N. Craig Smith, "The New Corporate Philanthropy," *Harvard Business Review May–June 1991 issue*, accessed at https://hbr.org/1994/05/the-new-corporate-philanthropy, accessed November 2015.

[101] Ibid.

[102] Ibid.

[103] Kasturi Rangan, Lisa Chase, and Sohel Karim, "The Truth About CSR," *Harvard Business Review January–February 2015*, p. 4.

[104] FedEX, "FedEx Pledges US$1 Million in Aid to Support Relief of Nepal Earthquake Disaster," *FedEX website*, May 8, 2015, http://about.van.fedex.com/newsroom/asia-english/fedex-pledges-us1-million-in-aid-to-support-relief-of-nepal-earthquake-disaster/, accessed November 2015.

[105] Ibid.

FedEx's efforts showcased the importance of not only alignment, but also partnership for efficient corporate impact in community health.

Partnership with community organizations: Corporations could rarely impact community health without ties to nonprofit or public entities. Cross-collaboration across sectors often gave a corporation a better understanding of the community and magnified program impacts.

For example, in an effort to aid the 2.5 billion people who did not have access to safe and hygienic sanitation, the Gates Foundation provided a grant to Kohler Co, a global leader in kitchen and bathroom technology, and a team from the California Institute of Technology to develop a solar-powered toilet.[106] The California Institute of Technology team designed a prototype toilet in 2012 that won the Bill & Melinda Gates Foundation's "Reinventing the Toilet Challenge." The toilet included a self-contained water purification and disinfection system, which reused water and did not require wastewater disposal. Kohler provided plumbing products and design expertise to the team, as well as technical support for a field trial in India.[107]

The effort highlighted the unique skills each group could bring to the initiative. In 2013, Rob Zimmerman, Kohler's sustainability marketing manager discussed the cross-collaboration efforts, stating:

> It is exciting and certainly an honor for us to work with the Caltech team, who are true pioneers of their time. Kohler is known for pioneering innovative products and helping to advance technology, and through the Gates Foundation challenge, we get the opportunity to support others in their efforts to push traditional systems to a new level.[108]

Coordination among different corporations: There were many examples of companies pursuing community health as part of their social responsibility activities, and many of these companies worked with non-profit or public entities to aide and scale their efforts. However, few companies worked directly with other companies to pool resources and combat shared community health challenges. Research had shown that corporate social responsibility (CSR) programs were rarely coordinated internally at corporations[109]—let alone between different corporations. Working together on CSR programs could be more efficient than working alone.[110] Furthermore, communities stood to benefit from the joint, rather than fragmented efforts.

---

[106] Kohler Co., "Kohler Co. Supports Caltech's Solar-Powered Toilet Project," *PR Newswire*, November 19, 2013, http://www.prnewswire.com/news-releases/kohler-co-supports-caltechs-solar-powered-toilet-project-232553791.html, accessed May 2016.

[107] Ibid.

[108] Ibid.

[109] Kasturi Rangan, Lisa Chase, and Sohel Karim, "The Truth About CSR," *Harvard Business Review January–February 2015.*

[110] Tim Mohin, "The Top Trends in CSR for 2012," *Forbes*, January 18, 2012, accessed at http://www.forbes.com/sites/forbesleadershipforum/2012/01/18/the-top-10-trends-in-csr-for-2012/#2715e4857a0b147a902824a4, accessed January 2016.

Some corporations had taken advantage of these efficiencies to combat community health challenges. Two of the largest corporations in Maine, General Dynamics Bath Iron Works ("BIW") and L.L. Bean, teamed up to promote diabetes prevention in their employees, dependents, and community.[111] Though the corporations had very different businesses—BIW was a ship manufacturer and defense contractor and L.L. Bean was a retail company—the companies realized that many of their employees had spouses or other family members working at the other organization and that diabetes was a common challenge.[112] Working together, they jointly held diabetes prevention workshops in community YMCAs and other locations.[113]

Measurement protocols: Measurement protocols before program implementation were useful for understanding community needs, while measurements taken after implementation helped to determine program success. One community's challenges were not always the same as another's, and corporations interested in involvement in community health first had to understand the health challenges in that community.

In this realm, for-profit companies could learn from the non-profit hospital sector. Dignity Health, a faith-based provider organization, developed the first standardized Community Need Index (CNI) that enabled it to quantify community health risk across the nation.[114] It based a community's CNI score on five underlying socioeconomic barriers that affected overall health: income, cultural/language, educational, insurance, and housing.[115] (See Exhibit 4.4 for a comparison of a CNI scores in high-need and low-need communities.) Dignity Health found that CNI scores were positively correlated with hospital utilization, so it used the scores to develop programs and services aimed at the underlying causes of health in an effort to improve health and reduce healthcare utilization.[116]

---

[111] Vera Oziransky, Derek Yach, Tsu-Yu Tsao, Alexandra Luterek, Denise Stevens, "Beyond the Four Walls: Why Community Is Critical to Workforce Health," The Vitality Institute, July 2015, accessed at http://thevitalityinstitute.org/site/wp-content/uploads/2015/07/VitalityInstitute-BeyondTheFourWalls-Report-28July2015.pdf, accessed November 2015.

[112] Ibid.

[113] Kathleen Pierce, "Working in Wellness: Bath Ironworks and L.L. Bean," *The Maine Magazine*, April 2015, accessed at https://www.themainemag.com/play/wellness/2703-working-in-wellness-bath-ironworks-and-llbean.html, accessed January 2016.

[114] Dignity Health, "Improving Public Health & Preventing Chronic Disease: Dignity Health's Community Need Index," accessed at https://www.dignityhealth.org/stjosephs/about-us/community-benefit/community-building/documents/dignity-health-community-need-index-brochure, accessed December 2015.

[115] Ibid.

[116] Ibid.

Exhibit 4.4: Comparison of CNI Scores for High-Need and Low-Need Communities

		Green Valley, AZ 85614		Compton, CA 90220	
Barrier	Indicator	Indicator %	Barrier Score	Indicator %	Barrier Score
Income	Elderly Poverty Child Poverty Single Parent Poverty	3% 8% 32%	3	17% 27% 40%	4
Cultural	Non-Caucasian Limited English	8% 1%	2	97% 16%	5
Education	Without HS Diploma	9%	1	45%	5
Insurance	Unemployed Uninsured	4% 13%	2	15% 32%	5
Housing	Renting %	12%	1	38%	4
Final CNI Score			1.8 (Low Need)		4.6 (High Need)

Source: Dignity Health, "Improving Public Health & Preventing Chronic Disease: Dignity Health's Community Need Index," accessed at https://www.dignityhealth.org/stjosephs/about-us/community-benefit/community-building/documents/dignity-health-community-need-index-brochure, accessed December 2015.

The benefits of the program were multiple. It enabled Dignity Health to improve the health of the individuals it served, and because the program aimed to reduce utilization, it had the potential to reduce health system costs. Further, though many nonprofit hospitals and health systems already conducted community needs assessments prior to the Patient Protection and Affordable Care Act (ACA), they were formally required to do so following its passage, making tools like the CNI even more important for hospitals.[117]

In addition to measuring community needs prior to interventions, measuring the outcomes was essential for understanding whether a program was successful or not. Despite this, very few corporations had extensively measured outcomes from their community health programs, and when they did, they more often measured the impact on their own employees rather than the entire community. However, measuring the success of community health programs was challenging—not only for corporations, but also for public health practitioners, such as community organizations and medical providers.[118] By 2015, measurement protocols and tools for community health interventions were discussed in the public health literature, but it remained a developing area. On this topic, a 2015 National Academy of Medicine paper stated:

> Community development projects that improve housing conditions, public safety, employment, transportation, walkability, and access to green space and healthy food can have a profound impact on health outcomes. In most cases, however, practitioners of community

[117] IRS, "New Requirements for 501©(3) Hospitals Under the Affordable Care Act," accessed at https://www.irs.gov/Charities-&-Non-Profits/Charitable-Organizations/New-Requirements-for-501(c)(3)-Hospitals-Under-the-Affordable-Care-Act, accessed December 2015.

[118] Maggie Super Church, "Using data to address health disparities and drive investment in healthy neighborhoods," December 3, 2015, *National Academy of Medicine*, accessed at http://nam.edu/perspectives-2015-using-data-to-address-health-disparities-and-drive-investment-in-healthy-neighborhoods/, accessed January 2016.

development and medicine do not have the ability to measure the impact of these projects over time. As a result, the health benefits and related cost savings of these interventions remain essentially invisible.[119]

Conclusion

The relationship between a community's circumstances and an individual's health had been firmly established—and this had significant consequences for corporations.[120,121] It was increasingly clear that businesses could not simply operate *in* a community, they needed to be *of* the community—both understand and respond to the unique challenges their communities faced. Corporations, in the long run, were far more successful financially when they were embedded in healthy communities, as improved community health spawned healthier employees and consumers. Moreover, community health was influenced by a wide range of socioeconomic factors; simply improving access to health care did not automatically create healthier communities.

This note has assessed the drivers behind corporate investments in community health, the long history between corporations and communities, as well as the different ways in which corporations impacted community health. Renewed interest in the role of corporations in community health suggested that the field was still evolving, though several questions about the future remained.

Discussion Questions on Community Health

1. **Should corporations take leadership roles in community health promotion?** **Or, should this responsibility fall to non-profit and/or public actors instead?**

   (a) By 2015, many corporations provided community health support—whether it be for fundamental public health services, such as access to clean water, or for more ancillary services, such as raising awareness around public health issues. Are these appropriate corporate activities? What are the risks for corporations that invest in community health? What are the risks for communities that depend on corporations for these services?

[119] Ibid.

[120] Harry J. Heiman and Samantha Artiga, "Beyond Health Care: The Role of Social Determinants in Promoting Health and Health Equity," *The Kaiser Family Foundation*, November 4, 2015, accessed at http://kff.org/disparities-policy/issue-brief/beyond-health-care-the-role-of-social-determinants-in-promoting-health-and-health-equity, accessed November 2015.

[121] The World Health Organization, Commission on Social Determinants of Health, "closing the gap in a generation: Health equity through action on the social determinants of health," 2008, http://apps.who.int/iris/bitstream/10665/43943/1/9789241563703_eng.pdf, accessed November 2015.

2. **Many corporations, like Humana, make large investments in community health and gain a competitive advantage through the subsequent improvements in community health (e.g., healthier employees or consumers). How should corporations that make large investments in community health think about the potential problem of moral hazard**?

    (a) Humana and its work in San Antonio was given in this note as an example of exceptional community heath work. In that case, when community health improves, both the community and Humana see benefits. In a hypothetical situation, a second insurer—who hasn't invested in community health—enters the San Antonio market. That insurer takes advantage of Humana's community health work by offering lower priced health insurance products to the now-healthier community members. How should corporations think about this type of challenge? How can they protect themselves competitively?

3. **How can corporations foster trusted, high-impact partnerships for pursuing community health**?

    (a) From local alliances with government to agreements with nonprofits that spanned continents, the role of partnership in community health was highlighted time and time again. Businesses rarely impacted communities on their own. How can corporations convene the right community stakeholders for maximum long-term impact?
    (b) Though many corporations sought out public and non-profit partners, there were fewer examples of corporations working with other corporations to pursue community health programs. Corporations in the same geographic area may be working towards similar community health goals, yet they pursue programs in a fragmented way through their disparate corporate social responsibility programs. Is it feasible for companies to pool their resource and coordinate with each other to pursue community health initiatives? How can corporations seek out these kinds of partnerships?

4. **What are the best ways to measure corporate impact in community health**?

    (a) As discussed, outcome measurement is critical for understanding community health program success. However, measurement tools are often difficult to design and remain underutilized. What are the most appropriate measurement tools that corporations can use to understand the impact of their community health efforts? Which variables are most important to track?

    (b) If your corporation has previously implemented a community health program or initiative, have you succeeded or failed using any type of measurement?

Chapter 5
Environmental Health

In November 2015, world leaders from over 190 countries gathered in Paris at the 2015 Paris Climate Conference ("COP21") to discuss challenges posed by climate change, as well as potential solutions. Between 1900 and the mid-2000s, the Earth warmed by 0.7 °C.[1] Though climate projection models varied, temperature increases were expected to continue.[2] Furthermore, scientific evidence had mounted throughout the 2000s that these changes in the Earth's environment were largely due to human activities.[3] In its 2014 synthesis report, the Intergovernmental Panel on Climate Change (IPCC), an international body, provided its view on climate change effects[4]:

> Human influence has been detected in warming of the atmosphere and the ocean, in changes in the global water cycle, in reductions in snow and ice, and in global mean sea level rise; and it is extremely likely to have been the dominant cause of the observed warming since the mid-twentieth century. In recent decades, changes in climate have caused impacts on natural and human systems on all continents and across the oceans.[5]

---

[1] N. H. Stern, "The Economics of Climate Change: The Stern Review," *Cambridge University Press, Cambridge, UK*, 2007.

[2] United States Environmental Protection Agency (EPA), "Future Climate Change," *EPA website,* accessed at http://www3.epa.gov/climatechange/science/future.html#ref2, accessed January 2016.

[3] The Core Writing Team, R.K. Pachauri and L.A. Meyer, IPCC, 2014: "Climate Change 2014: Synthesis Report. Contribution of Working Groups I, II and III to the Fifth Assessment Report of the Intergovernmental Panel on Climate Change, *IPCC,* Geneva, Switzerland, p. 47, accessed at http://epic.awi.de/37530/1/IPCC_AR5_SYR_Final.pdf, accessed January 2016.

[4] Intergovernmental Panel on Climate Change (IPCC), "Organization," *IPCC website,* accessed at http://www.ipcc.ch/organization/organization.shtml, accessed January 2016.

[5] The Core Writing Team, R.K. Pachauri and L.A. Meyer, IPCC, 2014: "Climate Change 2014: Synthesis Report. Contribution of Working Groups I, II and III to the Fifth Assessment Report of the Intergovernmental Panel on Climate Change, *IPCC,* Geneva, Switzerland, p. 47, accessed at http://epic.awi.de/37530/1/IPCC_AR5_SYR_Final.pdf, accessed January 2016.

J.A. Quelch, E.C. Boudreau, *Building a Culture of Health*, SpringerBriefs in Public Health, DOI 10.1007/978-3-319-43723-1_5

Though greenhouse gases were released as a result of natural processes, many human activities also increased greenhouse gas emissions.[6] Greenhouse gases included carbon dioxide, methane, nitrous oxide, and fluorinated gases.[7] These gases accumulated and trapped heat in the atmosphere by altering the incoming solar radiation and outgoing infrared radiation.[8] Carbon dioxide (CO_2), was the most commonly discussed and greatest threat, as it had caused more than half of the total amount of warming produced by greenhouse gas emissions from human activities,[9] such as the burning of fossil fuels for transportation, heating and cooling, and manufacturing among other activities.[10,11]

Researchers had found that it was likely that the impact of these climate changes would be widespread—affecting not only the natural environment, but also the health and economic welfare of individuals and communities.[12] Therefore, the outcomes of COP21 were highly anticipated—though warily, as 5 years prior a similar climate change meeting in Copenhagen fell short of expectations when the attending countries could not agree on a deal.[13]

Unlike Copenhagen, world leaders in Paris were able to reach an agreement. Ban Ki-moon, the United Nations secretary general said, "This is truly a historic moment. For the first time, we have a truly universal agreement on climate change, one of the

---

[6] S. Solomon, D. Qin, M. Manning, Z. Chen, M. Marquis, K.B. Averyt, M.Tignor and H.L. Miller, IPCC, 2007: "Climate Change 2007: The Physical Science Basis. Contribution of Working Group I to the Fourth Assessment Report of the Intergovernmental Panel on Climate Change," Cambridge University Press, Cambridge, United Kingdom and New York, NY, USA, accessed at http://oceanservice.noaa.gov/education/pd/climate/factsheets/howhuman.pdf, accessed January 2016.

[7] United States Environmental Protection Agency (EPA), "A Student's Guide to Climate Change: Greenhouse Gases," *EPA website,* accessed at http://www3.epa.gov/climatechange/kids/basics/today/greenhouse-gases.html, accessed January 2016.

[8] S. Solomon, D. Qin, M. Manning, Z. Chen, M. Marquis, K.B. Averyt, M.Tignor and H.L. Miller, IPCC, 2007: "Climate Change 2007: The Physical Science Basis. Contribution of Working Group I to the Fourth Assessment Report of the Intergovernmental Panel on Climate Change," Cambridge University Press, Cambridge, United Kingdom and New York, NY, USA, accessed at http://oceanservice.noaa.gov/education/pd/climate/factsheets/howhuman.pdf, accessed January 2016.

[9] United States Environmental Protection Agency (EPA), "A Student's Guide to Climate Change: Greenhouse Gases," *EPA website,* accessed at http://www3.epa.gov/climatechange/kids/basics/today/greenhouse-gases.html, accessed January 2016.

[10] S. Solomon, D. Qin, M. Manning, Z. Chen, M. Marquis, K.B. Averyt, M.Tignor and H.L. Miller, IPCC, 2007: "Climate Change 2007: The Physical Science Basis. Contribution of Working Group I to the Fourth Assessment Report of the Intergovernmental Panel on Climate Change," Cambridge University Press, Cambridge, United Kingdom and New York, NY, USA, accessed at http://oceanservice.noaa.gov/education/pd/climate/factsheets/howhuman.pdf, accessed January 2016.

[11] United States Environmental Protection Agency (EPA), "Sources of Greenhouse Gas Emissions," *EPA website,* accessed at http://www3.epa.gov/climatechange/ghgemissions/sources.html, accessed January 2016.

[12] N. H. Stern, "The Economics of Climate Change: The Stern Review," *Cambridge University Press, Cambridge, UK,* 2007.

[13] Coral Davenport, "Nations Approve Landmark Climate Accord in Paris," *The New York Times,* December 12, 2015, accessed at http://www.nytimes.com/2015/12/13/world/europe/climate-change-accord-paris.html?_r=0, accessed January 2016.

most crucial problems on earth."[14] The Center for Climate and Energy Solutions, an independent US nonprofit, outlined the outcomes of the conference, stating in a summary report:

"Together, the Paris Agreement and the accompanying COP decision:

- Reaffirm the goal of limiting global temperature increase well below 2 °C, while urging efforts to limit the increase to 1.5°
- Establish binding commitments by all parties to make 'nationally determined contributions' (NDCs), and to pursue domestic measures aimed at achieving them
- Commit all countries to report regularly on their emissions and 'progress made in implementing and achieving' their NDCs, and to undergo international review
- Commit all countries to submit new NDCs every 5 years, with the clear expectation that they will 'represent a progression' beyond previous ones
- Reaffirm the binding obligations of developed countries under the UNFCCC to support the efforts of developing countries, while for the first time encouraging voluntary contributions by developing countries too
- Extend the current goal of mobilizing $100 billion a year in support by 2020 through 2025, with a new, higher goal to be set for the period after 2025
- Extend a mechanism to address 'loss and damage' resulting from climate change, which explicitly will not 'involve or provide a basis for any liability or compensation'
- Require parties engaging in international emissions trading to avoid 'double counting'
- Call for a new mechanism, similar to the Clean Development Mechanism under the Kyoto Protocol, enabling emission reductions in one country to be counted toward another country's NDC."[15]

Corporate Involvement

For many years, private sector concern for environmental health had focused on a fairly narrow set of issues (e.g., avoiding chemical accidents, reducing toxic exposure risks for employees and communities).[16] However, by 2016, it was increasingly clear that the private sector had a role to play in broader environmental issues

[14] Ibid.

[15] Center for Climate and Energy Solutions, "OUTCOMES OF THE U.N. CLIMATE CHANGE CONFERENCE IN PARIS, December, 2015, accessed at http://www.c2es.org/international/negotiations/cop21-paris/summary, pdf download at http://www.c2es.org/docUploads/cop-21-paris-summary-12-2015-final.pdf, accessed January 2016.

[16] OECD, "THE ENVIRONMENT, HEALTH AND SAFETY PROGRAMME: Managing Chemicals through OECD," *OECD*, 2009–2012, accessed at http://www.oecd.org/env/ehs/1900785.pdf, accessed February 2016.

that also had the potential to impact health. These included the health effects of climate change, water pollution, species destruction, and waste disposal, among others.[17,18]

While much of the focus at COP21 was on the collaborative efforts of governments, the private sector was also present at the climate talks. In the aftermath of the conference, The World Bank discussed the "integral part" that the private sector played in the talks and would continue to play.[19] It stated:

> CEOs from industries as far ranging as cement, transportation, energy, and consumer producers stepped up their efforts to address climate change, making their own commitments to decrease their carbon footprints, adopt renewable energy and engage in sustainable resource management. Global financial institutions pledged to make hundreds of billions of investment available over the next 15 years for clean energy and energy efficiency. Throughout the negotiations, the private sector called on governments to put in place predictable, long-term regulatory regimes, including a price on carbon and incentives for decarbonization. They also encouraged adoption of supportive policies that could facilitate the transition.[20]

Some corporations had already taken significant steps to combat climate challenges and other environmental hazards. In 2015, Unilever, a consumer packaged goods company, vowed to become 'carbon positive' by 2030.[21] Unilever explained the definition of "carbon positive," stating that it meant that "100 % of our energy across our operations will come from renewable sources, and with partners we will directly support the generation of more renewable energy than we need for our own operations, making the surplus available to the markets and communities in which we operate."[22]

This initiative built upon Unilever's previous efforts to positively impact the environment. In 2010, the company unveiled the Unilever Sustainable Living Plan (USLP), which stated three main goals it wished to achieve by 2020: (1) to help a billion people improve their health and well-being; (2) to halve the environmental footprint of making and using Unilever products and; (3) to enhance livelihoods of those in its value chain.[23] (See Exhibit 5.1.)

---

[17] The World Health Organization (WHO), "Media centre: Climate change and health," *WHO website,* accessed at http://www.who.int/mediacentre/factsheets/fs266/en/, accessed January 2016.

[18] Annette Prüss, David Kay, Lorna Fewtrell, and Jamie Bartram, "Estimating the Burden of Disease from Water, Sanitation, and Hygiene at a Global Level," *Environmental Health Perspectives,* May 2002, Volume 110 Number 5, pp. 537–542.

[19] The World Bank, "Private Sector—an Integral part of Climate Action Post-Paris," *The World Bank,* December 30, 2015, http://www.worldbank.org/en/news/feature/2015/12/30/private-sector-an-integral-part-of-climate-action-post-paris, accessed January 2016.

[20] Ibid.

[21] Unilever, "Unilever to become 'carbon positive' by 2030," *Unilever website,* November, 27, 2015, accessed at https://www.unilever.com/news/news-and-features/2015/Unilever-to-become-carbon-positive-by-2030.html, accessed January 2016.

[22] Ibid.

[23] Christopher Bartlett, "Unilever's New Global Strategy: Competing through Sustainability," HBS No. 916-414 (Boston: Harvard Business School Publishing, 2014).

Exhibit 5.1: Unilever Sustainable Living Plan

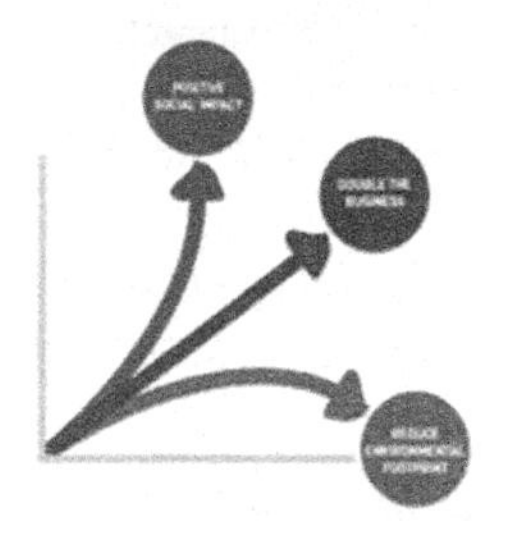

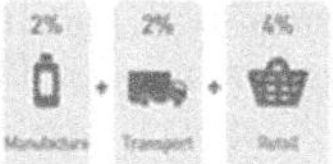

Source: USLP Summary of Progress 2014 ("Scaling for Impact"), p. 5.

At the same time, the company planned to double the size of its business.[24] A 2015 Harvard Business School case on Unilever stated:

> What made USLP unusual was its breadth. The commitment not only applied to every Unilever business, function, and country under its direct control, it also extended across its value chain and over the product lifecycle. This ambitious scope was revealed in an analysis at that time showing that Unilever's own manufacturing activities generated less than 5 % of its products' total greenhouse gas (GHG) footprint. Its suppliers contributed 21 % and consumers of its products accounted for 70 %. Accepting responsibility to halve that entire footprint represented a huge undertaking.[25]

While Unilever's sustainability programs were lauded for their expansive and aggressive goals, there were still many barriers to overcome. The company was dependent on actions from suppliers, consumers, and governments to achieve many of its objectives.[26] Further, critics had noted that some challenges persisted globally; specifically, the company had received criticism that poor working conditions continued at Unilever tea estates in Kenya and that the company failed to clean up a contaminated factory in India in a timely manner.[27] Unilever's experience highlighted the difficulties many corporations faced when taking largescale action against environmental challenges.

---

[24] Ibid.

[25] Ibid.

[26] David Gelles, "Unilever Finds That Shrinking Its Footprint Is a Giant Task," *The New York Times,* November 21, 2015, accessed at http://www.nytimes.com/2015/11/22/business/unilever-finds-that-shrinking-its-footprint-is-a-giant-task.html, accessed January 2016.

[27] Ibid.

The Environment and Health

Environmental conditions and public health were linked.[28] The World Health Organization (WHO) defined "environmental health" as follows:

> Environmental health addresses all the physical, chemical, and biological factors external to a person, and all the related factors impacting behaviours. It encompasses the assessment and control of those environmental factors that can potentially affect health. It is targeted towards preventing disease and creating health-supportive environments. This definition excludes behaviour not related to environment, as well as behaviour related to the social and cultural environment, and genetics.[29]

Climate change in particular was predicted to affect the environmental, social, and economic determinants of health.[30] A 2014 White House report, "The Health Impacts of Climate Change on Americans," stated:

> Climate change, caused primarily by carbon pollution, threatens the health and well-being of Americans in many ways, from increasing the risk of asthma attacks and other respiratory illnesses to changing the spread of certain vector-borne diseases. Some of these health impacts are already underway in the United States and climate change will, absent other changes, amplify some of the existing health threats the Nation now faces.[31]

Changing environmental conditions posed many hazards. Growing ground-level ozone was expected to increase asthma prevalence, while more extreme heat waves amplified the risk of hospitalization for individuals with cardiovascular, respiratory, and cerebrovascular diseases.[32] Particle pollution and smoke exposure from wildfires were also likely to increase respiratory and cardiovascular hospitalizations, as well as episodes of asthma, bronchitis, chest pain and respiratory infections.[33] Changing rainfall patterns were predicted to have different effects in different places—creating droughts in some areas and floods in others.[34] Water scarcity could affect food production and create famine, while floods could contaminate freshwater, heighten the risk of infection from water- or insect-borne diseases, and lead to

---

[28] Healthy People 2020, "Environmental Health," *HealthyPeople.gov website,* http://www.healthypeople.gov/2020/topics-objectives/topic/environmental-health, accessed December 2015.

[29] The World Health Organization (WHO), "Environmental health," *WHO website,* accessed at http://www.who.int/topics/environmental_health/en/, accessed January 2016.

[30] The World Health Organization (WHO), "Media centre: Climate change and health," *WHO website,* accessed at http://www.who.int/mediacentre/factsheets/fs266/en/, accessed January 2016.

[31] The White House, "The Health Impacts of Climate Change on Americans," June 2014, accessed at https://www.whitehouse.gov/the-press-office/2014/06/06/white-house-releases-report-health-impacts-climate-change-americans, PDF download at https://www.whitehouse.gov/sites/default/files/docs/the_health_impacts_of_climate_change_on_americans_final.pdf, accessed January 2016.

[32] Ibid.

[33] Ibid.

[34] The World Health Organization (WHO), "Media centre: Climate change and health," *WHO website,* accessed at http://www.who.int/mediacentre/factsheets/fs266/en/, accessed January 2016.

Table 5.1 Estimated costs of extreme weather events in the US (2012)

Event	Cost
Drought/Heatwave	$30B
Superstorm Sandy	$65B
Western Wildfires	$1B
Hurricane Isaac	$2.3B
Other severe weather incidents	$11.1B

Source: The White House, "Climate Change and President Obama's Action Plan," *The White House website,* accessed at https://www.whitehouse.gov/climate-change, accessed February 2016

drownings.[35] Environmental conditions were additionally expected to change patterns of infection (e.g., the prevalence of malaria and dengue fever were highly affected by climate).[36]

In addition to the hazards posed by the environmental conditions, climate change also had the potential to impact health in more indirect ways by affecting economic and social conditions. Put simply, extreme weather events were economically disastrous (see Table 5.1 for estimated costs of extreme weather events in 2012 in the US.)

Economic costs of such weather events fell to governments, individuals, and businesses. The Stern Review, a widely cited report on the economic effects of climate change, found that both rich and poor countries were likely to feel the detrimental economic effects of climate change.[37] The report stated:

> The evidence shows that ignoring climate change will eventually damage economic growth. Our actions over the coming few decades could create risks of major disruption to economic and social activity, later in this century and in the next, on a scale similar to those associated with the great wars and the economic depression of the first half of the twentieth century. And it will be difficult or impossible to reverse these changes. Tackling climate change is the pro-growth strategy for the longer term, and it can be done in a way that does not cap the aspirations for growth of rich or poor countries. The earlier effective action is taken, the less costly it will be.[38]

Though climate change was one of the most significant threats to environmental health in 2016, it was not the only one. As the earlier definition indicated, environmental health included all physical, chemical, and biological factors that impacted health.[39] This encompassed air and water contamination from both indoor and outdoor toxins, as well as challenges like the loss of biodiversity—and therefore genetic diversity necessary for new innovative medicines—due to the destruction of rainforests.[40]

---

[35] Ibid.

[36] Ibid.

[37] N. H. Stern, "The Economics of Climate Change: The Stern Review," *Cambridge University Press, Cambridge, UK*, 2007.

[38] Ibid.

[39] The World Health Organization (WHO), "Environmental health," *WHO website,* accessed at http://www.who.int/topics/environmental_health/en/, accessed January 2016.

[40] The World Health Organization (WHO), "Biodiversity," *WHO website,* accessed at http://www.who.int/globalchange/ecosystems/biodiversity/en/, accessed February 2016.

Chemical pollution was particularly challenging. Though it was not a new issue in 2016, it remained a relevant environmental challenge for the private sector. While many chemicals—including those that occurred naturally and those that resulted from human activities—were a part of everyday life and not hazardous, some had the potential to cause significant damage to the environment and human health.[41] One review on the burden of disease due to chemicals stated:

> ...some [chemicals] are life-threatening on contact and some persist in the environment, accumulate in the food chain, travel large distances from where they are released, and are harmful to human health in small amounts. Human exposure can occur at different stages of the life-cycle of a chemical, including through occupational exposure during manufacture, use and disposal, consumer exposure, exposure to contaminated products, or environmental exposure to toxic waste. Exposure can occur via various pathways, including inhalation of contaminated air and dust, ingestion of contaminated water and food, dermal exposure to chemical or contaminated products, or fetal exposure during pregnancy.[42]

That review found that the largest chemical contributors to death and disability were indoor smoke from solid fuel use, outdoor air pollution, and second-hand smoke.[43] These were followed by occupational particulates, chemicals involved in acute poisonings, and pesticides involved in self-poisonings.[44] Although regulation had increased around corporate chemical use since the 1970s, many challenges remained.[45] The environment clearly had a large impact on human health in a variety of different ways, and corporations were involved in many activities that contributed to these challenges.

Issues in Environmental Health

Politics

In the US, climate change was a highly politicized topic, making regulatory action against it more difficult. Many Republican leaders questioned or denied the science behind climate change, and often criticized climate change policies put forth by

[41] Annette Prüss-Ustün, Carolyn Vickers, Pascal Haefliger, and Roberto Bertollini, "Knowns and unknowns on burden of disease due to chemicals: a systematic review," *Environ Health.*, 2011; 10: 9, accessed at http://www.ncbi.nlm.nih.gov/pmc/articles/PMC3037292/, accessed February 2016.

[42] Annette Prüss-Ustün, Carolyn Vickers, Pascal Haefliger, and Roberto Bertollini, "Knowns and unknowns on burden of disease due to chemicals: a systematic review," *Environ Health.*, 2011; 10: 9, accessed at http://www.ncbi.nlm.nih.gov/pmc/articles/PMC3037292/, accessed February 2016.

[43] Ibid.

[44] Ibid.

[45] OECD, "THE ENVIRONMENT, HEALTH AND SAFETY PROGRAMME: Managing Chemicals through OECD," *OECD,* 2009–2012, accessed at http://www.oecd.org/env/ehs/1900785.pdf, accessed February 2016.

Democratic leaders.[46] In September, 2015, Republican presidential candidate, Marco Rubio, said, "We are not going to make America a harder place to create jobs in order to pursue policies that will do absolutely nothing, nothing to change our climate."[47]

However, Democrats pushed for larger-scale action, and expressed frustration with naysayers. In his 2016 State of the Union address, President Obama not only criticized climate change skeptics, but also positioned climate change as a business opportunity, stating:

> Look, if anybody still wants to dispute the science around climate change, have at it. You'll be pretty lonely, because you'll be debating our military, most of America's business leaders, the majority of the American people, almost the entire scientific community, and 200 nations around the world who agree it's a problem and intend to solve it. But even if the planet wasn't at stake; even if 2014 wasn't the warmest year on record—until 2015 turned out even hotter—why would we want to pass up the chance for American businesses to produce and sell the energy of the future?[48]

Vulnerability

Not all people or communities were equally vulnerable to environmental health threats. Some countries were more vulnerable to climate change challenges due to their geographic location (e.g., coastal, island, and polar regions) or their lack of infrastructure to respond to such challenges (e.g., lack of medical facilities or disaster relief resources). Some developing countries faced both challenges. The Germanwatch Climate Risk Index analyzed the impacts of weather-related events in different countries around the world and provided a ranking of countries.[49] Its 2016 index stated, "of the ten most affected countries (1995–2014), nine were developing countries in the low income or lower-middle income country group, while only one was classified as an upper-middle income country."[50]

---

[46] Coral Davenport, "Nations Approve Landmark Climate Accord in Paris," *The New York Times,* December 12, 2015, accessed at http://www.nytimes.com/2015/12/13/world/europe/climate-change-accord-paris.html?_r=0, accessed January 2016.

[47] Staff writer, "Presidential Candidates on Climate Change," *The New York Times,* Election 2016, http://www.nytimes.com/interactive/2016/us/elections/climate-change.html?_r=0, accessed May 2016.

[48] The White House Office of the Press Secretary, "Remarks of President Barack Obama—State of the Union Address As Delivered," January 13, 2016, accessed at https://www.whitehouse.gov/the-press-office/2016/01/12/remarks-president-barack-obama-%E2%80%93-prepared-delivery-state-union-address, accessed January 2016.

[49] Sönke Kreft, David Eckstein, Lukas Dorsch & Livia Fischer, "GLOBAL CLIMATE RISK INDEX 2016," *Germanwatch e.V,* November 2015, PDF accessed at https://germanwatch.org/en/download/13503.pdf, accessed January 2016.

[50] Ibid.

President Obama recognized these issues in his remarks at the first session of COP21. He stated:

> We know the truth that many nations have contributed little to climate change but will be the first to feel its most destructive effects. For some, particularly island nations—whose leaders I'll meet with tomorrow—climate change is a threat to their very existence. And that's why today, in concert with other nations, America confirms our strong and ongoing commitment to the Least Developed Countries Fund. And tomorrow, we'll pledge new contributions to risk insurance initiatives that help vulnerable populations rebuild stronger after climate-related disasters.[51]

However, issues of environmental health vulnerability went beyond climate change. Some communities were also more vulnerable to environmental challenges due to poor maintenance of and underinvestment in public infrastructure. In 2016, Flint, Michigan made national headlines for significant lead contamination in its drinking water.[52] The crisis was caused by aging pipes that had been poorly maintained. Ingesting lead could cause damage to the brain, kidneys, nerves and blood; it could also cause behavioral problems, learning disabilities, seizures and death.[53] One *Fortune* article, noting that the concerns around public infrastructure were much broader than Flint, Michigan, stated:

> Flint, Michigan is certainly not the first US city to see its water contaminated by aging pipes. And, unless the many American cities with aging lead pipes get to work quickly, it will not be the last, either. The poisoned tap water of Flint serves as a warning sign to city officials throughout the US: Lead pipes installed in cities 100 or more years ago need to be replaced.[54]

Responsibility

By 2016, environmental threats to health were a global issue, but deciding who could—and should—take responsibility for them was an ongoing moral and political debate. Though responsibility in environmental health was often discussed from the perspective of climate change responsibility, it was also an issue for other environmental challenges. For example, the issue of chemical pollutant responsibility

[51] The White House Office of the Press Secretary, "Remarks by President Obama at the First Session of COP21," November 30, 2015, https://www.whitehouse.gov/the-press-office/2015/11/30/remarks-president-obama-first-session-cop21, accessed January 2016.

[52] Rob Curran, "Flint's Water Crisis Should Raise Alarms for America's Aging Cities," *Fortune,* January 25, 2016, http://fortune.com/2016/01/25/flint-water-crisis-america-aging-cities-lead-pipes/, accessed May 2016.

[53] U.S. Department of Housing and Urban Development, "About Lead-Based Paint," *HUD.gov website,* http://portal.hud.gov/hudportal/HUD?src=/program_offices/healthy_homes/healthyhomes/lead, accessed December 2015.

[54] Rob Curran, "Flint's Water Crisis Should Raise Alarms for America's Aging Cities," *Fortune,* January 25, 2016, http://fortune.com/2016/01/25/flint-water-crisis-america-aging-cities-lead-pipes/, accessed May 2016.

might be complicated by a complex, long, and global supply chain. The effort to determine responsibility amongst the countries, corporations, and individuals involved could be arduous.

The question of responsibility was particularly difficult among nations, as many developed countries had unknowingly contributed to modern environmental problems since the time of the industrial revolution—the 1700s.[55] Yet many of the countries that were most likely to experience the negative effects of climate change were developing countries that had not benefitted from the same level of long-term economic growth.[56] In 2012, UN Secretary General Ban Ki-moon stated, "The climate change phenomenon has been caused by the industrialisation of the developed world. It's only fair and reasonable that the developed world should bear most of the responsibility."[57]

Complicating matters further, by the late 2000s, many developing countries had increased their level of carbon emissions as they grew their economies. World leaders were charged with finding an ethical solution that allowed developing countries to continue to grow their economies, while also stemming emissions. In 2014, China ranked first in terms of total carbon dioxide emissions.[58] Prior to COP21, many developing countries, like China and India, were not required to take action against greenhouse gas emissions.[59] However, the agreement that came out of COP21 called for voluntary action from all countries.[60]

While the issue of responsibility was often discussed from a country-perspective, others had considered corporate responsibility. Not all corporations contributed to environmental issues in the same way. (See Exhibit 5.2 for the source of greenhouse gas emissions in 2012 by economic sector.) A 2014 study analyzed historic production records of the ninety largest producers of coal, oil, natural gas, and cement. It

---

[55] United States Environmental Protection Agency (EPA), "Causes of Climate Change," EPA website, accessed at http://www3.epa.gov/climatechange/science/causes.html#ref2, accessed January 2016.

[56] Sönke Kreft, David Eckstein, Lukas Dorsch & Livia Fischer, "GLOBAL CLIMATE RISK INDEX 2016," *Germanwatch e.V*, November 2015, PDF accessed at https://germanwatch.org/en/download/13503.pdf, accessed January 2016.

[57] Associated Press, "Ban Ki-moon: rich countries are to blame for global warming," *The Guardian,* December 5, 2012, accessed at http://www.theguardian.com/environment/2012/dec/05/ban-ki-moon-rich-countries, accessed January 2016.

[58] Jos G.J. Olivier, Greet Janssens-Maenhout, Marilena Muntean, Jeroen A.H.W. Peters, "Trends in global CO2 emissions: 2014 Report," *PBL Netherlands Environmental Assessment Agency Institute for Environment and Sustainability (IES) of the European Commission's Joint Research Centre (JRC),* PDF accessed at http://edgar.jrc.ec.europa.eu/news_docs/jrc-2014-trends-in-global-co2-emissions-2014-report-93171.pdf, accessed January 2016.

[59] Coral Davenport, "Nations Approve Landmark Climate Accord in Paris," *The New York Times,* December 12, 2015, accessed at http://www.nytimes.com/2015/12/13/world/europe/climate-change-accord-paris.html?_r=0, accessed January 2016.

[60] Center for Climate and Energy Solutions, "OUTCOMES OF THE U.N. CLIMATE CHANGE CONFERENCE IN PARIS, December, 2015, accessed at http://www.c2es.org/international/negotiations/cop21-paris/summary," pdf download at http://www.c2es.org/docUploads/cop-21-paris-summary-12-2015-final.pdf, accessed January 2016.

traced 63 % of cumulative worldwide emissions of industrial CO_2 and methane between 1751 and 2010 to the 90 "carbon major" entities it analyzed—meaning that a relatively small number of corporations was responsible for nearly two-thirds of historic CO_2 and methane emissions.[61] The study concluded:

> Most analyses to date…consider responsibility for climate change in terms of nation-states. Such analyses fit the framework of international law, insofar as treaties and conventions are based on agreements between nation states…Shifting the perspective from nation-states to corporate entities—both investor-owned and state-owned companies—opens new opportunities for those entities to become part of the solution rather than passive (and profitable) bystanders to continued climate disruption.[62]

Exhibit 5.2: Sources of U.S. Greenhouse Gas Emissions by Economic Sector

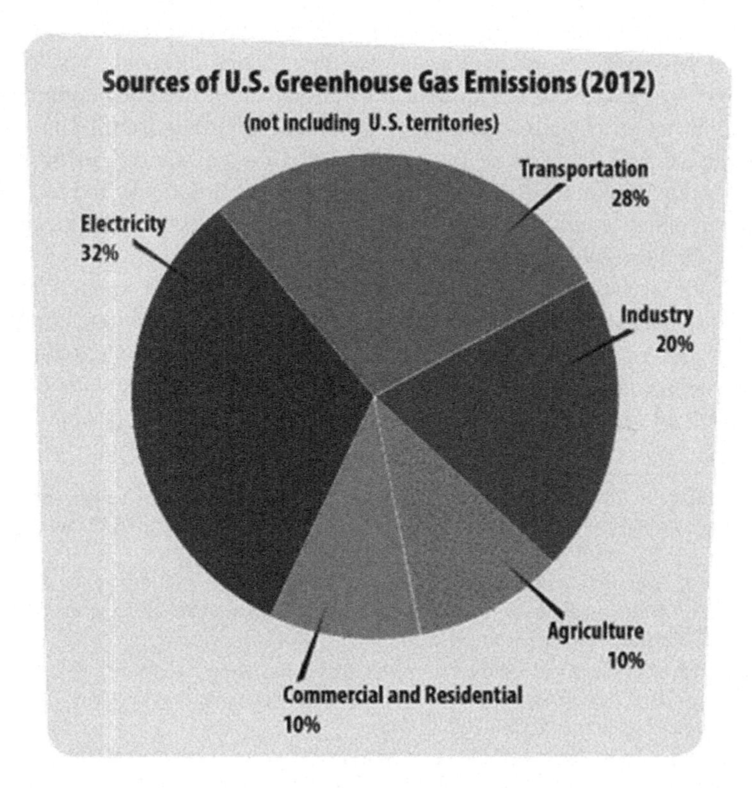

Source: United States Environmental Protection Agency (EPA), "A Student's Guide to Climate Change: Greenhouse Gases," *EPA website,* accessed at http://www3.epa.gov/climatechange/kids/basics/today/greenhouse-gases.html, accessed January 2016.

A 2015 study analyzed the responsibilities of the investor-owned fossil fuel producers specifically. The researchers concluded that those corporations should bear the responsibility for four main reasons:

---

[61] Richard Heede, "Tracing anthropogenic carbon dioxide and methane emissions to fossil fuel and cement producers, 1854–2010," *Climatic Change,* January 2014, Volume 122, Issue 1, pp. 229–241.

[62] Ibid.

1. "They have produced a large share of the products responsible for dangerous anthropogenic interference in the climate system.
2. They continued to produce them well after the danger was scientifically established and recognized by international policymakers.
3. They have worked systematically to prevent the political action that might have stabilized or reduced GHG emissions, including through unethical practices such as promoting disinformation.
4. While ostensibly acknowledging the threat represented by unabated reliance on fossil fuels, they nevertheless continue to engage in business practices that will lead to their expanded production and use for decades to come."[63]

Transparency

Neither countries nor corporations were always transparent about their environmental initiatives. Though sustainability reporting had increased throughout the 2000s, some barriers prohibited open reporting around environmental impact.[64] A 2015 incident involving Volkswagen ("VW"), a German automobile manufacturer, illustrated this point clearly. US regulators discovered that VW had installed defeat devices in its diesel car engines that enabled the cars to detect regulator testing conditions.[65] This allowed the engines to show a lower emissions result in a controlled, testing situation, but when the cars were used regularly by consumers, the engines emitted nitrogen oxide pollutants up to 40 times above US standards.[66]

What's more, even after regulators discovered the deception, VW remained evasive. In January, 2016, *The New York Times* reported that the company refused to share key documents with regulators.[67] Investigators hoped to obtain emails and other communications among VW executives to understand which employees knew about or approved the decision. New York's attorney general, Eric T. Schneiderman, said "Our patience with Volkswagen is wearing thin. Volkswagen's cooperation with the states' investigation has been spotty—and frankly, more of the kind one expects from a company in denial than one seeking to leave behind a culture of

[63] Peter C. Frumhoff, Richard Heede, and Naomi Oreskes, "The climate responsibilities of industrial carbon producers," *Climatic Change, S*eptember 2015, Volume 132, Issue 2, pp. 157–171.

[64] EY in association with the Global Reporting Initiative, "Sustainability reporting—the time is now," 2013, accessed at http://www.ey.com/Publication/vwLUAssets/EY-Sustainability-reporting-the-time-is-now/$FILE/EY-Sustainability-reporting-the-time-is-now.pdf, accessed January 2016.

[65] Russell Hotten, "Volkswagen: The scandal explained," *BBC News,* December 10, 2015, accessed at http://www.bbc.com/news/business-34324772, accessed January 2016.

[66] Ibid.

[67] DANNY HAKIM and JACK EWING, "VW Refuses to Give American States Documents in Emissions Inquiries," *The New York Times,* January 8, 2016, accessed at http://www.nytimes.com/2016/01/09/business/vw-refuses-to-give-us-states-documents-in-emissions-inquiries.html, accessed January 2016.

admitted deception."[68] An article on the public health impact of the VW emissions scandal estimated that the excess emissions would cause 59 early deaths in the US.[69] They found a social cost of ~$450 m when monetizing premature mortality.[70]

Why Corporations Support Environmental Health

There were many reasons why corporations supported environmental health. These included:

Potential to reduce costs: It had been firmly established that pursuing environmental objectives often resulted in efficiency gains that could reduce costs.[71] To reduce their impact on the environment, many corporations focused on cutting their water and energy use, as well as producing less waste. While objectives like these supported the environment, they also often lead to cost reductions for corporations.[72]

Regulations: Many corporations were required to meet certain environmental standards put forth by regulatory bodies. In 2014, researchers with the Organisation for Economic Co-operation and Development (OECD) analyzed data on the strictness of environmental laws in 24 OECD countries from 1990 to 2012 to assess whether environmental policies affected productivity growth.[73] While policies in all of the countries had become stricter since the 1990s, they found that the strictest environmental policies occurred in Denmark and the Netherlands, while the most relaxed were in Greece and Ireland. France, Britain, the US, and Poland were around the average.[74] Although China, the greatest producer of carbon emissions worldwide, had previously been criticized for relatively weak environmental protection laws, in

[68] Ibid.

[69] Steven R H Barrett, Raymond L Speth, Sebastian D Eastham, Irene C Dedoussi, Akshay Ashok, Robert Malina, and David W Keith, "Impact of the Volkswagen emissions control defeat device on US public health," *2015 IOP Publishing Ltd*, Environmental Research Letters, Volume 10, Number 11.

[70] Steven R H Barrett, Raymond L Speth, Sebastian D Eastham, Irene C Dedoussi, Akshay Ashok, Robert Malina, and David W Keith, "Impact of the Volkswagen emissions control defeat device on US public health," *2015 IOP Publishing Ltd*, Environmental Research Letters, Volume 10, Number 11.

[71] McKinsey & Company, "The business of sustainability: McKinsey Global Survey results," McKinsey & Company website, October, 2011, accessed at http://www.mckinsey.com/insights/energy_resources_materials/the_business_of_sustainability_mckinsey_global_survey_results, accessed January 2016.

[72] Ibid.

[73] Silvia Albrizio, E. Botta, T. Koźluk and V. Zipperer, "DO ENVIRONMENTAL POLICIES MATTER FOR PRODUCTIVITY GROWTH? INSIGHTS FROM NEW CROSS-COUNTRY MEASURES OF ENVIRONMENTAL POLICIES", *OECD Economics Department Working Papers*, No. 1176, *OECD Publishing*, 2014.

[74] Ibid.

2014 amended its environmental protection laws for the first time in 25 years.[75] It discussed plans to punish polluters more strictly as concerns grew that growing pollutants might have serious health effects.[76]

Threats to business: Environmental challenges—and climate change in particular—posed considerable threats to business (e.g., scarcity of natural resources, challenges to agriculture, destruction of operational infrastructure and supply chains).[77] In 2014, the World Bank stated, "In corporate boardrooms and the offices of CEOs, climate change is a real and present danger. It threatens to disrupt the water supplies and supply chains of companies as diverse as Coca-Cola and ExxonMobil. Rising sea levels and more intense storms put their infrastructure at risk, and the costs will only get worse."[78]

Environmental challenges due to failing public infrastructure, as in Flint, Michigan, could hinder business productivity. Because of this, some business were motivated to pursue public-private partnerships to ensure the environmental health and safety of the communities where they operated and did business.

Consumer expectations: Many corporations pursued environmental objectives because some consumers expected them to. In 2014, Nielsen studied global online consumers across 60 countries and whether they were willing to pay more for products and services provided by companies with committed to positive social and environmental impact. Nielsen stated, "The propensity to buy socially responsible brands is strongest in Asia-Pacific (64 %), Latin America (63 %) and Middle East/Africa (63 %). The numbers for North America and Europe are 42 and 40 %, respectively."[79]

However, research by McKinsey & Company suggested that while consumers were willing to pay more for green products, their willingness decreased as the premium for the green product increased.[80] (See Exhibit 5.3 for consumers' willingness to pay a premium for green products by industry). McKinsey surveyed con-

[75] Bloomberg News, "China Takes On Pollution With Biggest Changes in 25 Years," *Bloomberg Business,* April 25, 2014, accessed at http://www.bloomberg.com/news/articles/2014-04-24/china-enacts-biggest-pollution-curbs-in-25-years, accessed January 2016.

[76] Ibid.

[77] Victor Lipman, "Is Climate Change The Biggest Long-Term Management Problem Facing Business?," *Forbes,* February 4, 2014, accessed at http://www.forbes.com/sites/victorlipman/2014/02/04/is-climate-change-the-biggest-long-term-management-problem-facing-business/#24e76d0d2fc2, accessed February 2016.

[78] The World Bank, "World Bank Group President: This Is the Year of Climate Action," *The World Bank website,* January 23, 2014, accessed at http://www.worldbank.org/en/news/feature/2014/01/23/davos-world-bank-president-carbon-pricing, accessed February 2016.

[79] Nielsen Press Room, "GLOBAL CONSUMERS ARE WILLING TO PUT THEIR MONEY WHERE THEIR HEART IS WHEN IT COMES TO GOODS AND SERVICES FROM COMPANIES COMMITTED TO SOCIAL RESPONSIBILITY," *Nielsen website,* June 17, 2014, http://www.nielsen.com/us/en/press-room/2014/global-consumers-are-willing-to-put-their-money-where-their-heart-is.html, accessed February 2016.

[80] Mehdi Miremadi, Christopher Musso, and Ulrich Weihe, "How much will consumers pay to go green?," *McKinsey Quarterly,* October, 2012, accessed at http://www.mckinsey.com/insights/manufacturing/how_much_will_consumers_pay_to_go_green, accessed February 2016.

sumers in the US and Europe about purchases in the automotive, building, electronics, furniture, and packaging industries, and more than 70 % of the respondents indicated that they would pay an additional 5 % for a green product if it met the same performance standards as a traditional alternative.[81] However, as prices increased for the green products, the consumers' willingness to pay a premium decreased in all industries.[82]

Exhibit 5.3: Consumers' Willingness to Pay a Premium for Green Products by Industry

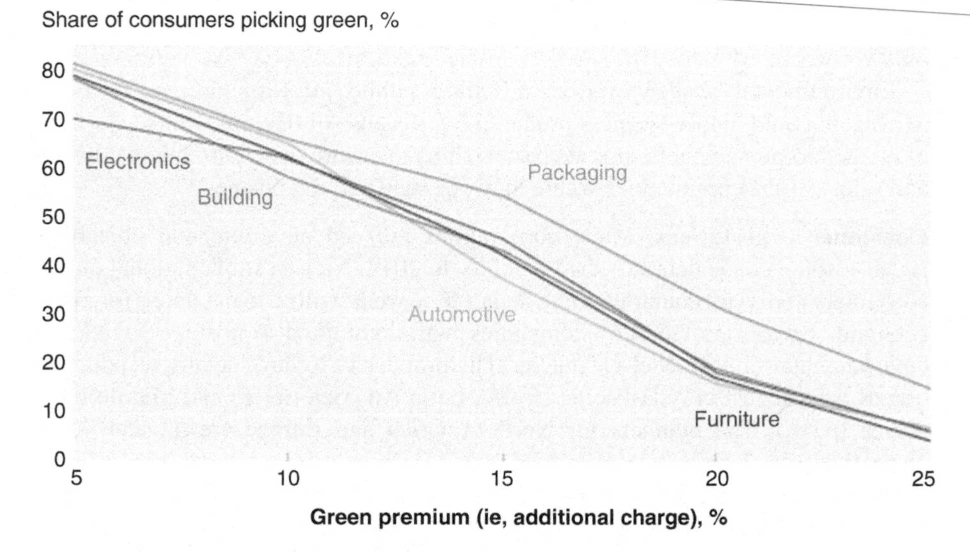

Source: "How much will consumers pay to go green?", October 2012, *McKinsey Quarterly*, www.mckinsey.com/mckinsey_quarterly. Copyright (c) 2012 McKinsey & Company. All rights reserved. Reprinted by permission.

Social responsibility: Corporate support of environmental safety and health was often also part of broader efforts to become more socially responsible. Corporations pursued social responsibility for a number of reasons, including out of moral obligation, for brand differentiation, and for consumer engagement.[83]

Barriers to Corporate Involvement in Environmental Health

Despite the motives behind corporate impact in environmental health, it was not always a top business priority for several reasons; these included:

---

[81] Ibid.

[82] Ibid.

[83] James Epstein-Reeves, "Six Reasons Companies Should Embrace CSR," *Forbes,* February 21, 2012, accessed at http://www.forbes.com/sites/csr/2012/02/21/six-reasons-companies-should-embrace-csr/, accessed November 2015.

Conflict with other business objectives: For some corporations—especially fossil fuel producers—promoting environmental health came into conflict with other business objectives. A 2015 article in the *Financial Times* noted that many energy companies continued to predict growth in traditional lines of business.[84] These companies were investing in new coal mines and oil explorations, which could be poor investments if new energy technology and further environmental policy reduced fossil fuel consumption.[85]

Lack of regulation: Regulation and enforcement of regulation were less rigorous in some countries. Even in countries where regulation was more robust, such as the US, critics noted that regulations could fall short. For example, in 2015, the U.S. Environmental Protection Agency (EPA) set new smog standards; however, the EPA was criticized for setting the standard too low.[86] A 2015 article in *Bloomberg* stated, "Smog causes or worsens asthma, heart disease and other ailments, especially among children. Yet the agency has chosen to impose the loosest standards it could, dismissing the warnings of its own scientific advisory panel. For a government otherwise committed to the environment, it's a remarkable misstep."[87] Although there was some concern that environmental regulation might harm growth, the 2014 OECD discounted this. It stated that "an increase in stringency of environmental policies does not harm productivity growth or productivity levels."[88]

Length of time for problems to appear: Although some environmental issues were apparent quickly, many, including climate change, were long term challenges. Individuals and businesses struggled to make the case for investments in environmental health as the returns could take decades.[89] Climate changes happened slowly; however, the economic effects of restricting emissions were felt in the present.[90]

---

[84] Andrew Hill, "COP21 Paris climate talks: Energy groups face Kodak moment," *Financial Times,* November 30, 2015, accessed at http://www.ft.com/cms/s/0/fb0c70dc-9515-11e5-ac15-0f7f7945adba.html#axzz3zDyjzuPM, accessed January 2016.

[85] Ibid.

[86] Editorial Board, "The EPA Chokes on Smog," *Bloomberg View,* October 1, 2015, accessed at http://www.bloombergview.com/articles/2015-10-01/the-epa-chokes-on-smog, accessed January 2016.

[87] Ibid.

[88] Silvia Albrizio, E. Botta, T. Koźluk and V. Zipperer, "DO ENVIRONMENTAL POLICIES MATTER FOR PRODUCTIVITY GROWTH? INSIGHTS FROM NEW CROSS-COUNTRY MEASURES OF ENVIRONMENTAL POLICIES", *OECD Economics Department Working Papers*, No. 1176, *OECD Publishing*, 2014, p. 28.

[89] Bryan Walsh, "Why We Don't Care About Saving Our Grandchildren From Climate Change," *TIME,* October 23, 2013, accessed at http://science.time.com/2013/10/21/why-we-dont-care-about-saving-our-grandchildren-from-climate-change/, accessed February 2016.

[90] Bryan Walsh, "Why We Don't Care About Saving Our Grandchildren From Climate Change," *TIME,* October 23, 2013, accessed at http://science.time.com/2013/10/21/why-we-dont-care-about-saving-our-grandchildren-from-climate-change/, accessed February 2016.

How Corporations Support Environmental Health

Leading businesses saw their environmental responsibilities extend across all aspects of the value chain. Corporate actions to advance environmental health broke down into six main categories. These categories were:

1. **Product and service innovation**: Companies enhanced product and service quality to advance environmental health.
2. **Operations management**: Businesses utilized innovative sourcing, manufacturing, and distribution strategies to reduce their carbon footprints and improve their use of raw materials. They also took actions to reduce chemical risks and effluents for employees and communities.
3. **Consumer use and disposal**: Businesses changed how consumers used their products and services and provided environmentally-friendly means of disposal.
4. **Philanthropic giving**: Corporations gave philanthropically to environmental causes.
5. **Investment strategies:** Businesses utilized investment strategies to promote better environmental practices and seek financial return.
6. **Policy engagement:** Companies sought ways to advance sustainable environmental policy.

Product and Service Innovation

Products

Many businesses invented or redesigned products to reduce negative impacts on environmental health. Companies achieved these goals in different ways—sometimes removing materials with known health risks (e.g., removing metals in textile dyes or phthalates in plastics) and other times, completely changing product design.[91,92] For example, Levi Strauss was one of the first companies to minimize the use of toxic chemicals in the manufacture of its apparel. It made its decision public, publishing a "Restricted Substances List" of chemicals that it would not permit.[93] Levi Strauss stated:

[91] Centers for Disease Control and Prevention (CDC), "Factsheet: Phthalates," *CDC website,* accessed at http://www.cdc.gov/biomonitoring/Phthalates_FactSheet.html, accessed February 2016.

[92] Tim Robinson, Geoff McMullan, Roger Marchant, Poonam Nigam, "Remediation of dyes in textile effluent: a critical review on current treatment technologies with a proposed alternative," *Bioresource Technology,* 2001 May; 77(3):247–55.

[93] Levi Strauss & Co, "Sustainability: Planet: Chemicals," *Levi Strauss & Co website,* http://www.levistrauss.com/sustainability/planet/chemicals/, accessed May 2016.

> From pesticides and fertilizers used in cotton production to dyes in the manufacturing process, chemicals touch the clothes we produce…Our Restricted Substances List (RSL) identifies the chemicals we will not permit in our products or in the production process due to their potential impact on consumers, workers and the environment. The RSL provides up-to-date information to our business partners, including direct sources and licensees, on product compliance with international consumer products regulations, to the best of our knowledge.[94]

Sometimes, a product's re-design was motivated by changing regulatory requirements, while other times, corporations were self-motivated to invent or redesign products. For example, many consumer products—from face washes to toothpastes—contained microbeads, small plastic beads added for their exfoliating abilities. They had become an environmental concern because they remained intact after consumers flushed them down the drain.[95] They were too small for water treatment plants to catch and filter out, and as a result, they had ended up in large numbers in many bodies of water. A 2015 study found that more than 8 trillion microbeads entered US bodies of water daily.[96] After fish, turtles and other wildlife ingested the beads, the plastic material would get caught in their digestive systems, blocking their absorption of important nutrients.[97]

Researchers did not fully understand the human health effects of these small plastic beads. Because of the chemicals in them and their longevity, there was concern that they could lead to cancer, cardiac problems, skeletal issues, endocrine disruption, or neurological deficiencies.[98] However, their use persisted in many products, and as a result, regulation was required to change industry behavior. At the end of 2015, President Obama approved a bill that banned the use of microbeads in cosmetic products; the new law required manufacturers to phase out the beads starting in July, 2017.[99]

Buildings traditionally used a large amount of energy—accounting for 73 % of electricity consumption in the US.[100] They also accounted for 38 % of CO_2 emissions and nearly 14 % of all potable water use.[101] Further, they typically lasted for a

---

[94] Ibid.

[95] Jareen Imam, "Microbead ban signed by President Obama," *CNN,* December 31, 2015, accessed at http://www.cnn.com/2015/12/30/health/obama-bans-microbeads/, accessed January 2016.

[96] Chelsea M. Rochman, Sara M. Kross, Jonathan B. Armstrong, Michael T. Bogan, Emily S. Darling, Stephanie J. Green, Ashley R. Smyth, and Diogo Veríssimo, "Scientific Evidence Supports a Ban on Microbeads," *American Chemical Society,* 2015, PDF accessed at http://pubs.acs.org/doi/pdf/10.1021/acs.est.5b03909, accessed January 2016.

[97] Christopher Johnson, "Personal Grooming Products May Be Harming Great Lakes Marine Life," *Scientific American,* June 25, 2013, http://www.scientificamerican.com/article/microplastic-pollution-in-the-great-lakes/, accessed January 2016.

[98] Ibid.

[99] Erin Brodwin, "Here's why the US government suddenly banned a bunch of soaps, bodywashes, and toothpastes," *Business Insider,* January 2, 2016, accessed at http://www.businessinsider.com/why-obama-banned-microbead-soap-2015-12, accessed January 2016.

[100] U.S. Green Building Council (USGBC), "Green Building Facts," *USGBC website,* February 23, 2015, accessed at http://www.usgbc.org/articles/green-building-facts, accessed January 2016.

[101] Ibid.

long period time, so their environmental impact was amplified over many years. In response, many construction firms vowed to create more efficient structures. Green buildings used less energy and water, and often minimized the environmental impact of construction. Researchers found that green buildings benefitted human health in two ways: "directly at the individual level through providing optimized indoor environments, and indirectly on a population level through reductions in energy use and thus reductions in air pollutants that cause premature death, cardiovascular disease, exacerbate asthma conditions and contribute to global climate change…"[102] "Green building" design became more common throughout the 2000s.[103] In 2015, 40–48 % of new nonresidential construction in the United States was green.[104]

Sometimes, the principles behind green building and design were also applied more broadly to neighborhood and urban design. For example, Vancouver, Canada aimed to become the greenest city in the world by 2020.[105] It set goals and targets to improve access to renewable energy, reduce waste, and increase the number of green buildings and the availability of green transportation. The city also hoped to increase residents' access to clean water and local food.[106] Though not led by the business community, the government in Vancouver engaged residents, private businesses, non-profit organizations, and all levels of government to implement its plan. Using this cross-sector approach, it explored innovative building codes, partnerships, and financing opportunities.[107]

Services

Companies were not only conscious of how their products affected the environment—they also considered the impacts of the services they delivered. Many corporations redesigned existing services, and others created new services that catered specifically to environmentally-conscious consumers. As the world's largest package delivery company, United Parcel Service ("UPS") had significant effects on the environment. The company depended on truck, train, and air travel to deliver packages—all activities that could potentially increase the company's carbon footprint.

[102] Joseph G. Allen, Piers MacNaughton, Jose Guillermo Cedeno Laurent, Skye S. Flanigan, Erika Sita Eitland, John D. Spengler, "Green Buildings and Health," *Current Environmental Health Reports,* September 2015, Volume 2, Issue 3, pp. 250–258.

[103] Ibid.

[104] U.S. Green Building Council (USGBC), "Green Building Facts," *USGBC website,* February 23, 2015, accessed at http://www.usgbc.org/articles/green-building-facts, accessed January 2016.

[105] City of Vancouver, "Greenest city goals," *City of Vancouver website,* http://vancouver.ca/green-vancouver/greenest-city-goals-targets.aspx, accessed May 2016.

[106] Ibid.

[107] City of Vancouver, "Sustainable programs for businesses and employees," *City of Vancouver website,* http://vancouver.ca/green-vancouver/sustainable-programs-for-businesses.aspx, accessed May 2016.

UPS stated, "Every day, UPS is faced with a complex challenge. How do we deliver more while using less?"[108]

To answer this question the company completely redesigned many aspects of its service to promote environmental health. It began tracking how many miles its trucks drove, and how much fuel, paper, and water it used.[109] UPS optimized route planning to minimize unnecessary travel, thus reducing greenhouse gas emissions. The company also began offering a carbon neutral shipping option. If a consumer chose this option, UPS purchased certified carbon offsets to balance out the emissions produced by the transportation of the consumer's shipment.[110] The company also tested a variety of new vehicles that used alternative fuels to advance research into new technologies.[111]

Other companies designed services for consumers with environmental health challenges. Ecolab, a specialty chemicals firm, provided cleaning and safety services to other businesses in foodservice, food processing, hospitality, healthcare, industrial, and oil and gas markets.[112,113] An article in *The Economist* stated, "Ecolab's research labs test deadly microbes from hospitals, plot ways to kill bed bugs more effectively and adjust chemical recipes to suit the water quality at customers' multiple locations. As hotel chains vie to offer the most comfortable beds, Ecolabbers investigate how to keep sheets soft after repeated washings."[114] While Ecolab enabled other companies to improve hygiene and keep their customers safe, some environmentalists criticized the organization for prioritizing cleanliness above "green" policies.[115]

---

[108] UPS, "Environmental Responsibility: Fuels & Fleets," *UPS website,* accessed at http://sustainability.ups.com/committed-to-more/fuels-and-fleets/, accessed January 2016.

[109] UPS, "Measurement Matters: UPS Carbon neutral," *UPS website,* https://www.ups.com/content/us/en/bussol/browse/carbon_neutral.html, accessed January 2016.

[110] Ibid.

[111] UPS, "Environmental Responsibility: Fuels & Fleets," *UPS website,* accessed at http://sustainability.ups.com/committed-to-more/fuels-and-fleets/, accessed January 2016.

[112] The Economist, "Cleaning up," *The Economist,* October 5, 2013, accessed at http://www.economist.com/news/business/21587277-bed-bugs-and-fracking-are-not-only-things-ecolab-has-going-it-cleaning-up, accessed January 2016.

[113] Ecolab, "About Us," *Ecolab website,* accessed at http://www.ecolab.com/about/, accessed January 2016.

[114] The Economist, "Cleaning up," *The Economist,* October 5, 2013, accessed at http://www.economist.com/news/business/21587277-bed-bugs-and-fracking-are-not-only-things-ecolab-has-going-it-cleaning-up, accessed January 2016.

[115] Ibid.

Operations Management

Sourcing

To improve their environmental impact, many corporations began to take a hard look at how they sourced their raw materials. In 2012, Ikea Group ("Ikea"), the world's largest home furnishings manufacturer, announced a goal of doubling revenues by 2020.[116] This target required the company to increase revenue by 10 % each year.[117] Ikea realized that this level of growth could have a tremendous impact on the environment. Around the same time that the organization announced these aggressive goals, it also launched a sustainability strategy called "People & Planet Positive." The program encompassed all aspects of Ikea's operations—from sourcing, to energy use, and waste creation among other things.[118] The organization explained the reasoning behind the venture, stating:

> By 2020, around 500 IKEA Group stores will welcome an estimated 1.5 billion visitors per year, employ more than 200,000 co-workers, potentially generating 45–50 billion euro in turnover. However, while that growth brings many great opportunities, if we continue with a business as usual approach, our use of wood will almost double and our carbon emissions will increase from today's 30 million tons to 50–60 million tons.[119]

The sourcing components of the sustainability program were heavily focused on Ikea's use of wood, as it was 60 % of the company's total raw material procurement by volume and 40 % by value.[120] However, sourcing sustainable wood was particularly challenging due to Ikea's growth plan. Ikea planned to expand into emerging markets, where wood supply chains were less sustainable, and the company often procured wood close to its consumer markets in an effort to minimize unnecessary costs.[121]

In its 2014 Sustainability Report, Ikea announced that it had increased the amount of wood it used from more sustainable sources from around 16 % in FY10 to over 41 % in FY14.[122] (See Exhibit 5.4 for Ikea's use of sustainably sourced wood from FY10 to FY14.) The company vowed to source 100 % of wood from more sustainable sources by 2020.[123] To achieve this goal, Ikea worked with global con-

---

[116] V. Kasturi Rangan, Michael W. Toffel, Vincent Dessain, Jerome Lenhardt, "Sustainability at IKEA Group," HBS No. 515-033 (Boston: Harvard Business School Publishing, 2015).

[117] Ibid.

[118] Ibid.

[119] IKEA Group, "People & Planet Positive IKEA Group Sustainability Strategy for 2020," *Ikea website,* PDF accessed at http://www.ikea.com/ms/en_GB/pdf/people_planet_positive/People_planet_positive.pdf, accessed January 2016.

[120] V. Kasturi Rangan, Michael W. Toffel, Vincent Dessain, Jerome Lenhardt, "Sustainability at IKEA Group," HBS No. 515-033 (Boston: Harvard Business School Publishing, 2015).

[121] Ibid.

[122] IKEA Group, "Sustainability Report FY14," PDF accessed at http://www.ikea.com/ms/en_US/pdf/sustainability_report/sustainability_report_2014.pdf, accessed January 2016.

[123] Ibid.

servation organizations, and supported existing wood suppliers to aide them in becoming more sustainable.[124] Ikea vowed to go beyond the immediate needs of its business and take a longer-term approach to forest management. It stated, "We want responsible forest management to be the norm, not just in the forests we use for our own products. We will contribute to ending deforestation by promoting the adoption of sustainable forestry methods across the whole industry, not just our business."[125]

Exhibit 5.4: IKEA's Use of Sustainably Sourced Wood (FY10 to FY14)

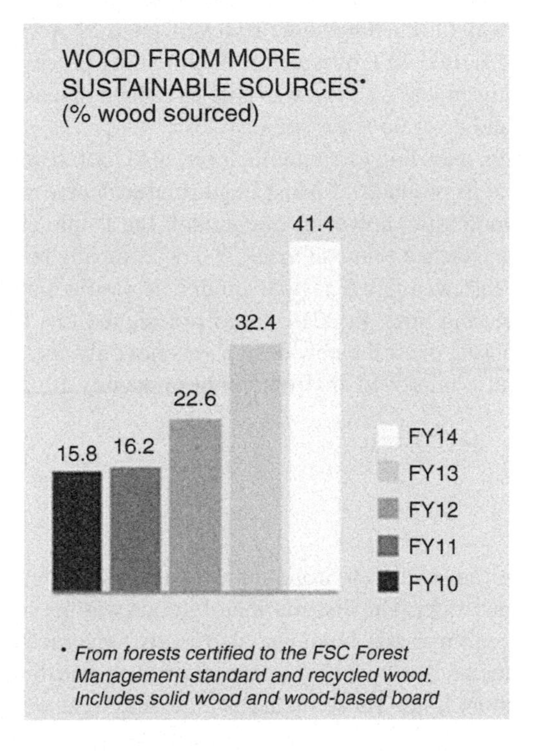

Source: IKEA Group, "Sustainability Report FY14," PDF accessed at http://www.ikea.com/ms/en_US/pdf/sustainability_report/sustainability_report_2014.pdf, accessed January 2016.

Manufacturing

Sourcing more sustainable materials was often a critical first step that corporations took to change their environmental impact; however, evaluating how those raw materials were then processed was as equally important. It was well-documented

---

[124] Ibid.

[125] Ibid.

that manufacturing contributed to negative environmental impacts through greenhouse gas emissions, as well as water and energy use. In the mid-2000s, researchers had found that despite a 70 % increase in the total manufacturing output, the total pollution emitted by US manufacturers had declined by about 60 % over previous 30 years.[126] This was due to technological innovation in manufacturing rather than changes in the mix of goods produced.[127]

In 2011, Levi Strauss & Co. ("Levi's"), a clothing company, dramatically reduced its environmental impact by reducing its use of water during manufacturing. Levi's developed a new way of finishing some styles of jeans.[128] According to Levi's, the average pair of jeans used 42 L of water in the finishing process; however, with the Water<Less™ jeans process, Levi's was able to reduce that amount of water by an average of 28 %, and up to 96 % for some styles.[129]

The Lego Group, a leading toy manufacturer, also took several steps to improve the sustainability of its products.[130] Most Lego products were made of one material, usually ABS thermoplastic; however, one model, the Duplo chassis, used an ABS thermoplastic base attached to metal axles.[131] The company realized that if it could use one material instead of two in that model, it would simplify assembly and decrease manufacturing time. In 2014, Lego redesigned this model, replacing the metal axles with plastic ones; the new design was not only cheaper for the company to produce, but it also improved environmental impact by 10–20 %.[132]

Distribution

Some corporations advanced environmental health by evaluating and redefining their distribution networks. The distribution of goods was not only a large contributor to greenhouse gas emissions, but it was also costly for companies.[133] Corporations used a variety of methods to reduce the impact of their distribution on the environment and their bottom lines. These included choosing light weight shipment pack-

[126] Levi Strauss, "Water<Less," *Levi Strauss Unzipped Blog,* November 3, 2010, accessed at http://www.levistrauss.com/unzipped-blog/2010/11/new-jeans-incredible-finishes-less-water/, accessed January 2016.

[127] Ibid.

[128] Ibid.

[129] Ibid.

[130] Maxine Perella, "Lego: how the signature brick is going green," *The Guardian,* April 11, 2014, accessed at http://www.theguardian.com/sustainable-business/lego-design-sustainability-circular-economy, accessed January 2016.

[131] Ibid.

[132] Ibid.

[133] Robert Duran Barbara, Neil McQuarrie, Elena Schrum, Terrence Teo, "Collaborative Distribution — An Analysis for the Environmental Defense Fund, *MIT Sloan School of Management,* PDF accessed at http://mitsloan.mit.edu/actionlearning/media/documents/s-lab-projects/EDF-report-2012.pdf, accessed January 2016.

aging, switching to distribution suppliers with climate change protocols, and using collaborative distribution when possible.[134]

In 2012, researchers from MIT Sloan School of Management analyzed collaborative distribution for the Environmental Defense Fund. They defined collaborative distribution as "the sharing of distribution resources between two or more shippers to send their goods within a common logistics network."[135] The authors pointed out that while the concept of collaborative distribution was not new in theory, the practice was not the norm.[136] Further, the practice was more common for goods shipped via air or sea, given the high cost, than for land transportation.[137]

In 2011, Ocean Spray and Tropicana, two competitors in the juice beverage market, formed a collaborative distribution network for rail shipments between Florida and New Jersey.[138] Ocean Spray had opened a new distribution center in Florida close to a rail yard already used by Tropicana.[139] Tropicana had been paying to send empty rail cars back from New Jersey to Florida, so Ocean Spray shifted 80 % of its shipping onto Tropicana's otherwise empty rail cars.[140] The switch reduced carbon emissions by 20 % and transportation costs by 40 %.[141]

Consumer Use and Disposal

Many businesses sought to change how consumers used and disposed of their products, as unnecessary waste could negatively impact the environment, and in turn, health.[142] In 2011, Patagonia, an outdoor clothing and gear company, created the Common Threads Initiative. It included a partnership between Patagonia and eBay to re-sell used items, as well as a program between Patagonia and its customers to

---

[134] Eric Lowitt, "How to Survive Climate Change and Still Run a Thriving Business," *Harvard Business Review,* April 2014.

[135] Robert Duran Barbara, Neil McQuarrie, Elena Schrum, Terrence Teo, "Collaborative Distribution—An Analysis for the Environmental Defense Fund," *MIT Sloan School of Management,* PDF accessed at http://mitsloan.mit.edu/actionlearning/media/documents/s-lab-projects/EDF-report-2012.pdf, accessed January 2016.

[136] Ibid.

[137] Ibid.

[138] Eric Lowitt, "How to Survive Climate Change and Still Run a Thriving Business," *Harvard Business Review,* April 2014.

[139] Justin Gerdes, "Shared shipping is slowly gaining ground between market rivals," *The Guardian,* August 11, 2014, accessed at http://www.theguardian.com/sustainable-business/2014/aug/11/collaboration-shipping-oceanspray-tropicana-hershey-pollution, accessed January 2016.

[140] Ibid.

[141] Ibid.

[142] Healthy People 2020, "Environmental Health," *HealthyPeople.gov website,* http://www.healthypeople.gov/2020/topics-objectives/topic/environmental-health, accessed December 2015.

encourage them to reconsider their purchase and disposal of products.[143] In 2011, Yvon Chouinard, Patagonia's founder and owner, said:

> The Common Threads Initiative addresses a significant part of today's environmental problem—the footprint of our stuff. This program first asks customers to not buy something if they don't need it. If they do need it, we ask that they buy what will last a long time—and to repair what breaks, reuse or resell whatever they don't wear any more. And, finally, recycle whatever's truly worn out. We are the first company to ask customers to take a formal pledge and be partners in the effort to reduce consumption and keep products out of the landfill or incinerator.[144]

On Black Friday in 2011, Patagonia ran a full page ad in *The New York Times* that stated, "Don't Buy This Jacket."[145] While many lauded the efforts, critics were quick to point out that the program did not necessarily dissuade consumers from buying new products.[146] One *Bloomberg* article stated in 2013:

> Not surprisingly, the corporate plea didn't work, which is to say it worked perfectly for a burgeoning company in the business of selling $700 parkas. In 2012—which included about nine months of the "buy less" marketing—Patagonia sales increased almost one-third, to $543 million, as the company opened 14 more stores. Last year, revenue ticked up another 6 percent, to $575 million. In short, the pitch helped crank out $158 million worth of new apparel.[147]

Philanthropic Giving

However, even critics noted that Patagonia's sustainability efforts were expansive and went well beyond the "buy less" campaign.[148] In addition to that campaign, Patagonia had pledged 1 % of its sales to "the preservation and restoration of the natural environment" since 1985.[149] It stated, "As a company that uses resources and produces waste, we recognize our impact on the environment and feel a

---

[143] Forest Reinhardt, Ramon Casadesus-Masanell, Lauren Barley, "Patagonia (B)," HBS No. 714-465 (Boston: Harvard Business School Publishing, 2014).

[144] PR Newswire, "Patagonia Launches Common Threads Initiative: A Partnership With Customers to Consume Less," *PR Newswire,* September 7, 2011, accessed at http://www.prnewswire.com/news-releases/patagonia-launches-common-threads-initiative-a-partnership-with-customers-to-consume-less-129372068.html, accessed January 2016.

[145] Forest Reinhardt, Ramon Casadesus-Masanell, Lauren Barley, "Patagonia (B)," HBS No. 714-465 (Boston: Harvard Business School Publishing, 2014).

[146] Kyle Stock, "Patagonia's 'Buy Less' Plea Spurs More Buying," *Bloomberg Business,* August 28, 2013, http://www.bloomberg.com/bw/articles/2013-08-28/patagonias-buy-less-plea-spurs-more-buying, accessed January 2016.

[147] Ibid.

[148] Ibid.

[149] Patagonia, "1 % for the Planet," *Patagonia website,* accessed at https://www.patagonia.com/us/patagonia.go?assetid=81218, accessed January 2016.

responsibility to give back."[150] Many corporations, like Patagonia, used philanthropic donations to impact the environment.

Some corporations even built philanthropic endeavors into their business models to support environmental health initiatives. While he was working on a consulting project in Africa, Peter Thum witnessed the severe health consequences many people endured due to poor access to clean water.[151] As a result, he developed Ethos® Water, a bottled water company that donated part of its profits for each bottle to clean water initiatives in developing countries. In 2005, he sold the company to Starbucks.[152] The program continued with Starbucks contributing $.05US ($.10CN in Canada) for each bottle sold.[153] Starbucks stated in 2016 that "more than $12.3 million has been granted to help support water, sanitation and hygiene education programs in water-stressed countries—benefiting more than 500,000 people around the world."[154]

Investment Strategies

While these kinds of philanthropic programs continued—many with great success—by the late 2000s, some corporations also began to use investment strategies to create both environmental and financial returns. According to the Climate Policy Initiative, the amount of climate finance invested globally was $391 billion in 2014.[155] Private sector investment accounted for $243 billion of this total, while the public sector contributed $148 billion—meaning that the private sector was the larger source of climate finance.[156] Further, private finance had increased from 2013 to 2014, driven by investments in renewable energy technologies.[157]

Founded in 2014, the Closed Loop Fund was a social impact fund focused on alleviating recycling challenges in the US.[158] Many recycling plants lacked the

---

[150] Ibid.

[151] Jessica Harris, "Ethics in a bottle: How Peter Thum broke the rules to build a charitable—and profitable—company," *CNN Money,* November 5, 2007, accessed at http://money.cnn.com/2007/10/31/smbusiness/Ethos.fsb/index.htm?postversion=2007110109, accessed January 2016.

[152] Ibid.

[153] Starbucks, "Ethos® Water Fund," *Starbucks website,* accessed at http://www.starbucks.com/responsibility/community/ethos-water-fund, accessed January 2016.

[154] Ibid.

[155] Barbara K. Buchner, Chiara Trabacchi, Federico Mazza, Dario Abramskiehn and David Wang, "Global Landscape of Climate Finance 2015," *Climate Policy Initiative,* November, 2015, accessed at http://climatepolicyinitiative.org/wp-content/uploads/2015/11/Global-Landscape-of-Climate-Finance-2015.pdf, accessed January 2016.

[156] Ibid.

[157] Ibid.

[158] David Gelles, "Big Companies Put Their Money Where the Trash Is," *The New York Times,* November 28, 2015, accessed at http://www.nytimes.com/2015/11/29/business/energy-environment/big-companies-put-their-money-where-the-trash-is.html, accessed January 2016.

infrastructure to sort and process the existing range of different plastic recyclables while still making a profit.[159] This meant that even when consumers recycled their used plastics, some of these products were never processed and were instead sent to landfills. Not only did local governments want their residents to recycle more frequently, as it led to reduced landfill waste and associated costs, but there was also a demand from many large companies for more recycled plastic.[160] Additionally, it had been documented for many years that recycling led to health benefits. More than 15 years prior to the advent of the Closed Loop Fund, The Environmental Protection Agency stated that "Recycling is a highly effective strategy for reducing all the categories of health risks and pollution resulting from virgin material production."[161]

To create change in the industry, the $100 million Closed Loop Fund provided both no-interest loans to cities and below-market-rate loans to companies that sought to build recycling infrastructure.[162] Closed Loop Fund investors included many large corporations, such as: Walmart, Coca-Cola, PepsiCo, Johnson & Johnson, Procter & Gamble, Unilever, Keurig Green Mountain, Goldman Sachs, 3 M and Colgate-Palmolive.[163] The fund had clear investment criteria for remedying recycling challenges. It stated:

> Every Closed Loop Fund investment in recycling infrastructure deliver measurable financial and environmental returns. First, it is a loan and it must pay back. Equally important, each investment must divert significant tonnage from landfill to the recycling stream. Leveraging pay-for-success as much as possible, each loan must have transparent reporting and clear line of site to repayment to allow for easy replication by other cities and companies.[164]

The investments were expected to create benefits for a variety of stakeholders. The fund itself would earn a 2 % management fee, cities and recycling companies would receive increased access to capital, and the corporate investors could expect their principal returned with interest.[165] Further, many of the corporate investors hoped to increase the available amount of recycled plastic for their use.[166]

---

[159] Ibid.

[160] Ibid.

[161] Environmental Protection Agency, "Recycling…for the future," November, 1998, accessed at https://ofee.gov/Resources/Guidance_reports/Guidance_reports_archives/future.pdf, accessed January 2016.

[162] David Gelles, "Big Companies Put Their Money Where the Trash Is," *The New York Times,* November 28, 2015, accessed at http://www.nytimes.com/2015/11/29/business/energy-environment/big-companies-put-their-money-where-the-trash-is.html, accessed January 2016.

[163] Closed Loop Fund, "About," *Closed Loop Fund website,* accessed at http://www.closedloopfund.com/about/, accessed January 2016.

[164] Ibid.

[165] David Gelles, "Big Companies Put Their Money Where the Trash Is," *The New York Times,* November 28, 2015, accessed at http://www.nytimes.com/2015/11/29/business/energy-environment/big-companies-put-their-money-where-the-trash-is.html, accessed January 2016.

[166] Ibid.

The Breakthrough Energy Coalition was founded in November, 2015.[167] The founders included Bill Gates, Marc Benioff, Mark Zuckerberg, and many others who were committed to finding new energy sources.[168] The group planned to invest in several sectors, including: the electricity generation and storage, transportation, industrial use, agriculture, and energy system efficiency.[169] The Breakthrough Energy Coalition website stated:

> The world needs widely available energy that is reliable, affordable and does not produce carbon. The only way to accomplish that goal is by developing new tools to power the world. That innovation will result from a dramatically scaled up public research pipeline linked to truly patient, flexible investments committed to developing the technologies that will create a new energy mix.[170]

Policy Engagement

Corporations also pursued environmental health objectives by advancing and supporting sustainable environmental policy. Many companies demanded greater clarity on environmental objectives from governments, and by 2016, there were several organizations in the US and Europe that vowed to organize and mobilize the private sector to push policy forward. In the US, one of the most influential organizations was Ceres, a non-profit organization founded by a small group of investors in the late 1980s in response to the Exxon Valdez oil spill.[171] The group described itself with the following statement:

> Ceres is an advocate for sustainability leadership. Ceres mobilizes a powerful network of investors, companies and public interest groups to accelerate and expand the adoption of sustainable business practices and solutions to build a healthy global economy. Our mission is to mobilize investor and business leadership to build a thriving, sustainable global economy.[172]

We Mean Business, a similar coalition of over 300 companies working towards more sustainable environmental policy, had "formed a common platform to amplify the business voice, catalyze bold climate action by all, and promote smart policy frameworks."[173] The group encouraged companies to pursue one or more of the following initiatives:

---

[167] Bill Tucker, "Just In Time For COP21; The Breakthrough Energy Coalition," *Forbes,* November, 30, 2015, accessed at http://www.forbes.com/sites/billtucker/2015/11/30/just-in-time-for-cop21-the-breakthrough-energy-coalition/#2715e4857a0b3e4e2df0223f, accessed January 2016.

[168] Ibid.

[169] Ibid.

[170] Breakthrough Energy Coalition, "Introducing the Breakthrough Energy Coalition," *Energy Coalition website,* accessed at http://www.breakthroughenergycoalition.com/en/index.html, accessed January 2016.

[171] Ceres, "Who we are," *Ceres website,* accessed at http://www.ceres.org/about-us/who-we-are, accessed February 2016.

[172] Ibid.

[173] We Mean Business, "About," *We Mean Business website,* accessed at http://www.wemeanbusinesscoalition.org/about, accessed February 2016.

- “Adopt a science-based emissions reduction target
- Put a price on carbon
- Procure 100 % of electricity from renewable sources
- Responsible corporate engagement in climate policy
- Report climate change information in mainstream reports as a fiduciary duty
- Remove commodity-drive deforestation from all supply-chains by 2020
- Reduce short-lived climate-pollutant emissions”[174]

Conclusion

By 2016, it was clear that environmental challenges threatened societal stability, business success, and individual health. This note has assessed the connections between environmental conditions and health, the drivers behind corporate actions in environmental health, as well as how some corporations have advanced sustainable objectives. However, many questions still remained.

Discussion Questions on Environmental Health

1. **Beyond legally mandated standards, what responsibility should corporations have to adopt and promote environmental health standards?**
   (a) How much of this responsibility should fall to governments or other actors?
   (b) Should some industries or specific organizations bear more responsibility for promoting environmental health?
   (c) What are the short and long-term corporate benefits associated with aggressive environmental policies? What are the risks?
2. **How can corporations and governments promote transparency and accountability around corporate actions that impact the environment, and in turn, health?**
   (a) As the Volkswagen emissions testing incident recently showcased, corporate transparency about environmental impact remains a challenge. Are there ways that corporations, industries, and/or governments can promote transparency around environmental health issues?
3. **What is the future role of private investment in environmental health?**
   (a) Although regulation have forced some industries to adopt more sustainable standards, in many others, private investors have helped to create new standards or programs. How will private funding play a role in the future?

[174] We Mean Business, “Take Action,” *We Mean Business website,* accessed at http://www.wemeanbusinesscoalition.org/take-action, accessed February 2016.

Chapter 6
Implementing a Culture of Health

All corporations intentionally and unintentionally impacted public health through their positive and negative contributions to consumer health, employee health, community health, and environmental health. The sum of these four impacts constituted a corporation's population health footprint (see Exhibit 6.1). Few corporations had yet calculated the sum of their combined impact on public health in this integrated way. However, by 2016, a handful of organizations were pursuing a Culture of Health—a culture in which health effects were consistently discussed and considered in everyday corporate decision-making. At these corporations, managers and employees strove to achieve as positive a population health footprint as possible. (See Exhibit 6.2 for sample population health footprint initiatives within three companies.)

Exhibit 6.1: Population Health Footprint (PHF)

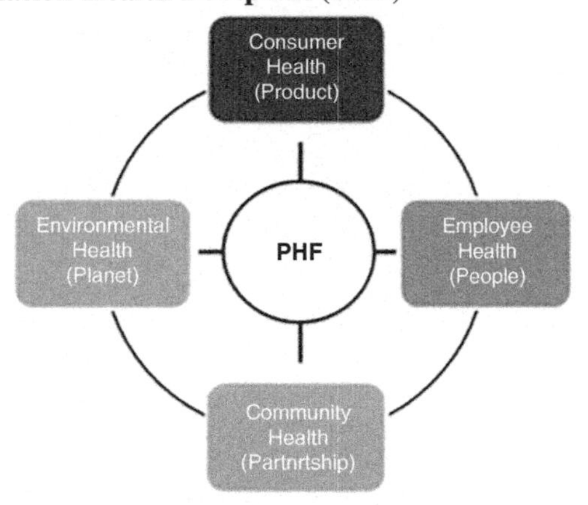

Source: Casewriter.

Reprinted with permission of Harvard Business School Publishing
Implementing a Culture of Health, HBS No. 9-516-112
This case was prepared by John A. Quelch and Emily C. Boudreau.

J.A. Quelch, E.C. Boudreau, *Building a Culture of Health*, SpringerBriefs in Public Health, DOI 10.1007/978-3-319-43723-1_6

Exhibit 6.2: Sample Population Health Footprint Initiatives for Three Companies

	Consumer health	Employee health	Community health	Environmental health
Royal Caribbean	– Created dedicated security teams on ships and at ports – Equipped ships to provide blood transfusions, mental health counseling, and medical imaging – Created an “outbreak prevention plan” to contain and treat norovirus, the leading cause of disease on cruise ships – Exceeded safety standards set by the Americans with Disabilities Act and the US Occupational Safety and Health Administration (OSHA)—based upon an Above and Beyond Compliance (ABC) philosophy	– Medically evaluated potential crew candidates; ensured that each was physically able to perform required duties – Offered crew medical checkups once every 2 years and access to yearly health fairs with free health screenings and educational resources (>90 % employee participation) – Offered employees on-ship and on-shore medical treatments – Offered mental wellness services with 24-hour assistance	– Promoted educational scholarship opportunities near Florida headquarters – Donated mattresses, towels, and furniture for reuse to community partners at ports-of-call, thereby reducing solid waste to landfills	– Managed wastewater, controlled solid waste, improved energy consumption, and reduced air emissions – Reduced greenhouse gas footprint by 19 % from 2005 to 2012 through a combination of reduced fuel use, purification systems that made emissions cleaner, and use of alternative energy sources
Target	– Introduced Made to Matter—Handpicked by Target, a collection of products from purpose-driven brands; made natural, organic, and sustainable products more accessible for consumers	– Offered medical, dental, vision and prescription drug coverage, as well as telehealth services and a NurseLine for 24/7 access to nurse consultation	– Founding member of the Alliance to Make US Healthiest, a coalition to help people become more physically and emotionally healthy	– Developed the Target Sustainable Product Index defined criteria for what makes a product more sustainable

	– Committed to increasing available organic food and beverages by 25 % by the end of 2017 – Introduced Simply Balanced grocery brand, which included over 40 % organic products; made it easier for consumers to find and choose organic produce in stores	– Offered tobacco cessation program, maternity support program for expecting mothers, well-being education resources, and health insurance incentive program – Offered employees a 20 % discount on fresh and frozen foods	– Offered employees opportunities to volunteer in their communities – Worked on the UNICEF Kid Power program; became the exclusive retailer of the Kid Power Band—"the world's first wearable-for-good,™ which helped supply nutritional food packets to poor kids around the world"[1]	– Committed to buying fish products from suppliers who sourced responsibly – Began program for making Target-brand packaging designs more sustainable by the end of 2016
Dow Chemical	– Given its risks as a chemical supplier, conducted product safety risk assessments through its "Business EH&S Risk Review (BRR) Work Process" – Retained third party verification partner to evaluate the robustness of Dow's product safety assessment processes	– Developed a diabetes program within its employee health promotion efforts, resulting in lower prevalence of chronic conditions among Dow employees (17 % lower than peer companies)	– Led the Michigan Health Improvement Alliance ("MiHIA"), a group of private, public, and non-profit partners that worked to improve health delivery and outcomes in central Michigan	- Employed over 100 scientists with expertise in mammalian toxicology, eco-toxicology, environmental science, industrial hygiene, human medicine and epidemiology to assess how Dow's businesses and chemical products affected the environment

Source: Casewriter analysis (Note: Entries for each company are intended to be illustrative, not comprehensive

[1] Target, "Wellness," *Target website*, https://corporate.target.com/corporate-responsibility/wellness, accessed April 2016.

By adopting a Culture of Health, corporations could reduce costs, increase revenues, and improve their reputations. However, this new mindset required an attitudinal shift from considering health as an expense to an investment, and the hard truth was that about 70 % of all corporate change initiatives failed.[2] To achieve a Culture of Health, corporations had to change internal mindsets, break down existing organizational silos, and promote corporate reporting standards that included measures of health impact.

Creating a Roadmap for Change

Any significant change management process required a roadmap showing the necessary steps to success—and creating a Culture of Health was no different. In the 1990s, John Kotter, of Harvard Business School, proposed a widely adopted eight-step process for leading change.[3] The process identified how corporations could transform quickly and where they were likely to fail. Table 6.1 applies Kotter's eight-step process to implementing a Culture of Health.

Rosabeth Kanter, of Harvard Business School, noted that lasting change often required multiple iterations and subtle improvements over time. She described her "Change Wheel," depicted in Exhibit 6.3, as follows[4]:

> In relatively small change efforts, one time around the Change Wheel might be enough to redirect the team, group, organization and to support progress toward the goal. Large-scale transformations, in contrast generally take many years and many turns of the wheel before the transformation is complete and sustainable. Change is fatiguing, and the temptation to stop (or declare "mission accomplished" while the war has barely started) is always present.[5]

[2] Nitin Nohria and Michael Beer, "Cracking the Code of Change," *Harvard Business Review*, May–June 2000 issue, accessed at https://hbr.org/2000/05/cracking-the-code-of-change, accessed April 2016.

[3] John P. Kotter, "Leading Change: Why Transformation Efforts Fail," *Harvard Business Review*, January, 2007, accessed at https://hbr.org/2007/01/leading-change-why-transformation-efforts-fail, accessed April 2016.

[4] Rosabeth Moss Kanter, "The Change Wheel: Elements of Systemic Change and How to get Change rolling," HBS No. 9-312-083 (Boston: Harvard Business School Publishing, 2011).

[5] Ibid.

Table 6.1 Kotter 8-stage change management process (applied to creating a culture of health)

Stage	Description	Culture of health actions needed
1	*Establish a sense of urgency*	• Examine untapped opportunities to improve health. Potential will differ by company (and by market, product/service, geographic area, company size, consumers, etc.)
		• Convince at least 75 % of managers that a larger focus on health is necessary
2	*Form a powerful guiding coalition*	• Break down silos between departments that currently address aspects of health separately (e.g. HR, product safety, sustainability, etc.)
		• Encourage managers from these departments to work as a team to focus on health
3	*Create a vision*	• Create a clear, compelling vision of the Culture of Health and explain its relevance to the overall mission and vision of the company
		• Develop strategies for realizing the Culture of Health vision throughout the organization and across all four areas of the population health footprint
4	*Communicate the vision*	• Communicate the Culture of Health vision frequently as part of the company's business plan
		• Suggest strategies for achieving the new health-focused vision
		• Use the guiding coalition to model and promote new health-focused decisions
5	*Empower others to act on the vision*	• Remove or alter systems/structures that undermine the Culture of Health vision
		• Encourage nontraditional ideas related to health-promoting activities
6	*Plan for and create short-term wins*	• Define, engineer, and showcase visible performance improvements related to one of more areas within the Culture of Health (e.g., those that offer significant, measurable cost savings)
		• Recognize and reward employees who contribute to those improvements and to the development of integrated Culture of Health solutions
7	*Consolidate improvements and produce more change*	• Use increased credibility from early wins to change systems, structures and policies that impede or undermine creating a Culture of Health
		• Hire, train, promote, and develop employees to understand, value, and can implement a Culture of Health
		• Develop new projects across different areas within the Culture of Health (i.e., if initially focused on employee health, begin a new project related to environmental health)
8	*Institutionalize new approaches*	• Ensure integrated strategy across the four pillars of the population health footprint
		• Articulate connections between new health-promoting activities and corporate revenues, profits, and reputation
		• Develop, measure, and refine the company's population health footprint
		• Share results widely, internally and externally

Source: John P. Kotter, "Leading Change: Why Transformation Efforts Fail," *Harvard Business Review*, January, 2007, https://hbr.org/2007/01/leading-change-why-transformation-efforts-fail, accessed April 2016; casewriter analysis

Exhibit 6.3: Change Wheel

Source: Adapted by casewriter from Rosabeth Moss Kanter, "The Change Wheel: Elements of Systemic Change and How to get Change Rolling," HBS No. 312-083 (Boston: Harvard Business School Publishing, 2011).

Corporate leaders could not build a Culture of Health overnight. They needed to assess their corporations' existing values and determine how to add or integrate "health" without overly disrupting the status quo or displacing existing values that were important or long-standing. Moving too quickly could have negative consequences. For example, if corporate leaders pushed for consumer health changes too quickly, consumers might receive hastily designed, second rate products. Claims of health as a corporate value would then risk being discredited.

Organizing for Change

Before plunging into the health care arena, any corporation had to research and understand the health needs of its stakeholders and assess the particular health-related initiatives that aligned with its resources and capabilities. It was important that any company's commitment not duplicate the efforts of others and not be so expensive as to be unattainable or likely to open the company to criticism. On the other hand, a company's Culture of Health vision should not be defined so narrowly as to be easily attainable and uninspiring.

Driving a Culture of Health through a corporation required the sustained commitment of the CEO; the appointment by the CEO of a former line manager with internal credibility to lead an integrated health strategy on a day-to-day basis; and changes in recruiting, training and incentive policies that emphasized the heightened importance of achieving a net positive population health footprint.

Vision, Values, and the Chief Executive

A corporate chief executive who committed to infusing a Culture of Health in his/her organization had to plan on at least a five-year journey to make health a core company value. The CEO had to see it through personally to make the shift in values stick. If the CEO saw the Culture of Health as part of his/her permanent legacy (in addition to good share price performance), so much the better. Focusing on health as a core value was always harder if it required displacing or subordinating commitment to other social endeavors, such as education, and always easier if there were important health-related initiatives already underway. The Culture of Health vision could serve as an exciting rallying point for employees in mundane industries who were not already engaged in social responsibility initiatives.

A company's CEO and senior leadership team were inherently role models for the rest of the organization. As such, they needed to exemplify a commitment to health in their own behaviors, regularly stress the importance and purpose of the Culture of Health vision, and deliver direction to drive the entire organization to consider the health impacts of all corporate decisions.[6] Cigna, a health insurance company, advocated for a corporate Culture of Health in a 2010 whitepaper:

> Leadership support from the top down is essential for changing beliefs and values. First and foremost, organizational leaders must both support—and show their support for—maintaining good health. This support needs to be heard from the CEO's own voice and shown through initiatives that cascade down through the ranks of the organization.[7]

The opportunity for the CEO to have great influence was particularly high in large retail and consumer-facing service organizations, like Enterprise Rent-A-Car and Target, where many impressionable young employees were in their first jobs after school or college.[8]

---

[6] Kevin Sharer, "How Should Your Leaders Behave?" *Harvard Business Review*, October 2013, https://hbr.org/2013/10/how-should-your-leaders-behave, accessed April 2016.

[7] Cigna, "Creating a Culture of Health," 2010, http://www.cigna.com/assets/docs/improving-health-and-productivity/837897_CultureOfHealthWP_v5.pdf, accessed May 2016.

[8] Tara Weiss, "Enterprise Offers That 'New Job' Smell," *Forbes*, March 27, 2008, accessed at http://www.forbes.com/2008/03/27/jobs-workforce-graduates-lead-careers-cx_tw_0327enterprise.html, accessed April 2016.

Operational Leadership

CEOs and senior leaders could provide vision for a Culture of Health, but a company also needed an operational leader to chair a multi-department Culture of Health task force and to oversee all health-related programs on a day-to-day basis—a Chief Health and Wellbeing Officer or a Chief Health and Sustainability Officer. However, in 2016, few organizations retained a single executive who was responsible for integrating and advancing all of a company's health activities, leading to a fragmented perspective.

Typically, product design and quality assurance managers attended to the healthfulness and safety of products and services sold to consumers. Human resource managers or chief medical officers covered employee health issues and designed employee wellness programs. Vice presidents of community relations or companies' foundations allocated funds to community health and other projects. Directors of sustainability reduced the company's carbon footprint and water use. These leaders were typically dispersed across different departments and reported to different executives.

For example, a chief medical officer (CMO), responsible for health services to employees, usually reported through human resources, while a chief sustainability officer (CSO) reported through manufacturing or external affairs, and in some cases, directly to the CEO.[9] Because of the different educational backgrounds traditionally required for these positions (medical for a CMO and business or engineering for a CSO), it was rare for one person to oversee both. At Siemens, for example, a single executive—who was also an executive director on the management board—was responsible for both sustainability and supply chain management, while a second executive served as CMO.[10,11]

CSOs were increasingly common; the number of companies with a full-time sustainability officer doubled between 1995 and 2003, and then doubled again between 2003 and 2008.[12] The title and role emerged first in the chemical industry where companies were under severe criticism for environmental pollution. As the definition of sustainability broadened from environmental impacts to include

---

[9] Kathleen Perkins Miller and George Serafeim, "Chief Sustainability Officers: Who Are They and What Do They Do?," Chapter 8 in *Leading Sustainable Change*, Oxford University Press, 2014, accessed at http://papers.ssrn.com/sol3/papers.cfm?abstract_id=2411976, accessed April 2016.

[10] Siemens, "Additional Sustainability information to the Siemens Annual Report 2013," *Siemens website*, accessed at http://www.siemens.com/about/sustainability/pool/en/current-reporting/siemens_ar2013_sustainability_information.pdf, accessed April 2016.

[11] Siemens, "Executive Biography," http://usa.healthcare.siemens.com/about-us/richard-frank, accessed May 2016.

[12] Kathleen Perkins Miller and George Serafeim, "Chief Sustainability Officers: Who Are They and What Do They Do?," Chapter 8 in *Leading Sustainable Change*, Oxford University Press, 2014, accessed at http://papers.ssrn.com/sol3/papers.cfm?abstract_id=2411976, accessed April 2016.

employee well-being throughout the supply chain, CSOs were increasingly required to expand their perspectives and assess products, planet, and people together. One study suggested that the role of the CSO evolved through three stages, from Compliance, through Efficiency, to Innovation:

> In terms of responsibilities we find that almost all CSOs in the first two stages (*Compliance* and *Efficiency*) perform a generic set of activities such as formulating and executing a sustainability strategy, identifying material sustainability issues, learning from external sources, reporting sustainability data, managing stakeholder relations and educating employees about sustainability. In contrast, in the *Innovation* stage we find a significantly lower frequency of CSOs engaging in most of those activities. We turn to our interview data to understand why and we find that this could be attributed to organizational needs becoming more idiosyncratic and CSOs decentralizing activities and decision rights.[13]

Both Royal Caribbean and Amazon could be considered in the "Innovation" stage. Each company showed)high levels of integration and each had a single person responsible for health, safety, security, and the environment, highlighting how health and wellbeing were increasingly integrated with sustainability. Royal Caribbean had a senior vice president of safety, environment, and health who reported directly to the CEO.[14] At Amazon, a single executive was responsible for health, safety, sustainability, security, and compliance. Each of these executives covered three of the four pillars within a Culture of Health (responsibility for community health was not included in either case). Though these integrated roles reflected management's realization of the interrelationships among the four pillars of the population health footprint, they remained largely focused on risk management, particularly in the case of Amazon.

The Royal Caribbean and Amazon executives could reasonably be considered Chief Health Officers given the breadth of their responsibilities. As companies adopted a Culture of Health, the title of "Chief Health and Sustainability Officer" was expected to become increasingly common. Appointees for these positions were likely to be drawn primarily from the sustainability and supply chain side of the business, with chief medical officers continuing in the organization structure but as one of their direct reports.

Incentives

Corporations also needed to identify line managers across different departments who were committed to the corporate Culture of Health agenda and needed to designate them as Culture of Health champions.[15] These executives would, in turn,

---

[13] Ibid.

[14] John A. Quelch and Margaret L. Rodriguez, "Royal Caribbean Cruises Ltd.: Safety, Environment, and Health," HBS No. 9-514-069 (Boston: Harvard Business School Publishing, 2015).

[15] John P. Kotter, "Leading Change: Why Transformation Efforts Fail," *Harvard Business Review*, January, 2007, accessed at https://hbr.org/2007/01/leading-change-why-transformation-efforts-fail, accessed April 2016.

motivate the employees in their units to embrace the Culture of Health vision. Part of their annual bonuses might depend on their effectiveness in doing so.

Corporations could use incentive systems to motivate managers to focus on corporate priorities. For example, at The Lego Group, manager bonuses were in a large part determined by each executive's contribution to advancing Lego's four promises: people (employee well-being), partner (with suppliers and distributors), planet (environmental health), and play (new product development to engage children and improve their cognitive development).[16]

While incenting Culture of Health leadership was critical, it was also important to motivate other employees to engage in this cultural shift and develop a shared understanding of purpose and desired outcomes. Corporations needed to ensure that building a Culture of Health was a grass roots activity, allowing employees, consumers and other stakeholders to surface innovative ideas for consideration. As one *Harvard Business Review* article that focused on managing change stated:

> The problem is that, while authority can compel action, it does little to inspire belief. It's not enough to get people to do what you want, they also have to want what you want—or any change is bound to be short-lived. That's why change management efforts commonly fail. All too often, they are designed to carry out initiatives that come from the top. ... To make change really happen, it doesn't need to be managed, but empowered. That's the difference between authority and leadership.[17]

As one example, Adidas recruited summer college interns for initial Climate Corps training and subsequent deployment to Adidas offices. They could submit sustainability project proposals to an internal venture capital fund with $3 million per year to allocate. The best projects were selected depending on their scope and impact and their return-on-investment projections.[18]

[16] Lego, "The LEGO GROUP Responsibility REPORT 2015," PDF downloaded from http://www.lego.com/en-us/aboutus/lego-group/annual-report, accessed April 2016.

[17] Greg Satell, "To Create Change, Leadership is More Important than Authority," *Harvard Business Review*, April 21, 2014, accessed at https://hbr.org/2014/04/to-create-change-leadership-is-more-important-than-authority/, accessed April 2016.

[18] Chris Martin, "The Greening of Adidas," *Bloomberg Businessweek*, May 5, 2016, accessed at http://www.bloomberg.com/news/articles/2016-05-05/adidas-saving-money-by-treating-energy-costs-like-vc-investments, accessed May 2016.

Networks of Support

Finally, corporate CEOs adopting the Culture of Health vision should not overlook engaging their corporate boards and foundations in the initiative. Corporations should have at least one health-knowledgeable director on their boards, and large companies also needed to align their foundations' objectives and funds with the Culture of Health initiative. The Lego Foundation, for example, which owned 25 % of the company, focused its resources (around one percent of Lego Group revenues) on providing children worldwide with safe and healthy places to play.[19]

Measuring Progress

Essential to institutionalizing a Culture of Health and persuading the skeptics was measuring and reporting corporate impacts across all four pillars of the population health footprint (consumer health, employee health, community health, and environmental health). As of 2016, there was no set framework integrating efforts across the four pillars into a simple net impact number. Research had shown that the current methods for reporting corporate social responsibility activity were not adequate for capturing a corporation's impact on public health.[20]

While there was no overall scorecard that covered the four pillars, considerable progress had been made in analyzing and reporting the impact of companies on both employee and environmental health. In the arena of employee health, The Vitality Institute, a research organization focused on health promotion, developed a three-part employee health scorecard. (See Exhibit 6.4 for the scorecard and Exhibit 6.5 for weights attached to each item measured.)

[19] Lego, "The LEGO GROUP Responsibility REPORT 2015," PDF downloaded from http://www.lego.com/en-us/aboutus/lego-group/annual-report, accessed April 2016.

[20] Rachel A. Spero, Fred D. Ledley, "Making Public Health Central to Standards for Corporate Social Responsibility," *Center for Integration of Science and Industry*: *Departments of Natural & Applied Science*, *Management*, Bentley University, 2015.

Exhibit 6.4: Vitality Institute Scorecard

GOVERNANCE - LEADERSHIP AND CULTURE	
1	Has your company conducted a confidential survey, audit, or other assessment within the present reporting period that measures the degree to which the workplace culture and environment support health and well-being? Examples: employees are asked to rate the corporate culture in some way; employees are asked if they feel their manager supports them when they take time to go to the gym at lunch
2	Are health, well-being, chronic disease prevention, or health promotion topics mentioned in - the annual report? - Form 10-K? - any other format reported to the board of directors at least once a year?
3	Is there a person responsible for employee health and well-being in your company?
MANAGEMENT - PROGRAMS, POLICIES, AND PRACTICES	
4	Does your company have an annual budget or receive dedicated funding for personalized health promotion and disease prevention programs? Examples: a dedicated budget in the department responsible for the implementation of the health and well-being program (e.g., the human resources department); a central health and well-being budget allocated by senior executives on an annual basis
5	Does your company have a program to address mental well-being, dealing with matters such as depression and stress management?
6	Does your company have an occupational safety and health (OSH) policy?
7	Does your company provide medical benefits for full-time workers, including recommended national preventive services (e.g., the Affordable Care Act in the United States) such as tobacco cessation, screenings, and vaccinations?
8	Does your company maintain a smoke-free workplace?
EVIDENCE OF SUCCESS - HEALTH RISKS AND HEALTH OUTCOMES	
9	Has your company conducted a confidential survey, audit, or other assessment within the present reporting period that measures the health status of employees?
10	What is the per-employee average absenteeism due to sick leave for the reporting period (unplanned leave or sick days)?

Source: Daniel Malan, Shahnaz Radjy, Nico Pronk, Derek Yach, "REPORTING ON HEALTH A Roadmap for Investors, Companies, and Reporting Platforms," *The Vitality Institute*, January 2016, accessed at http://thevitalityinstitute.org/site/wp-content/uploads/2016/01/Vitality-HealthMetricsReportingRoadmap22Jan2016.pdf, accessed April, 2016.

Exhibit 6.5: Vitality Institute Weighting System for Sustainability Scorecard

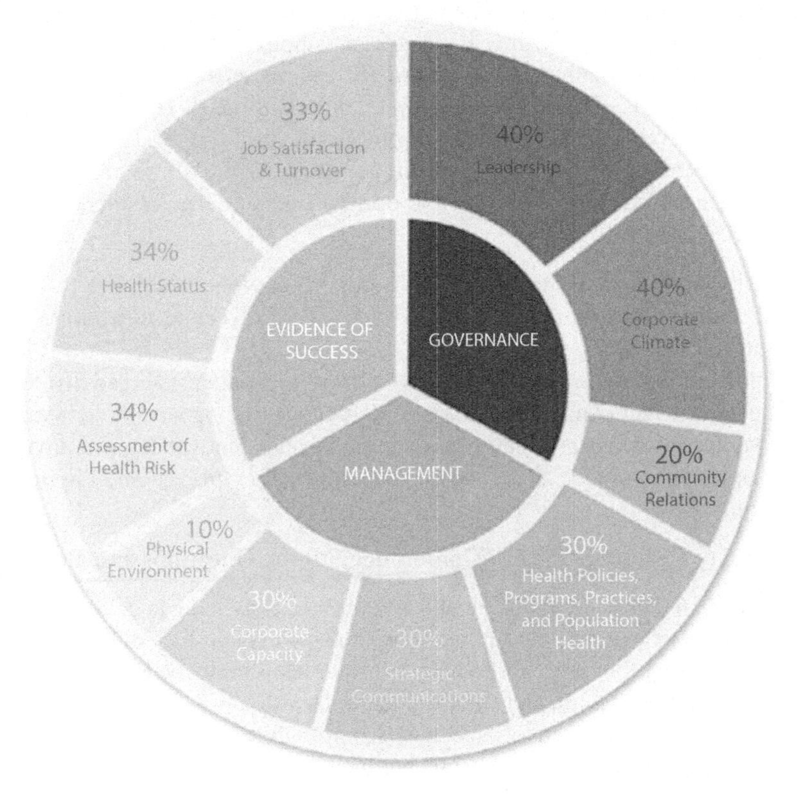

Source: Daniel Malan, Shahnaz Radjy, Nico Pronk, Derek Yach, "REPORTING ON HEALTH A Roadmap for Investors, Companies, and Reporting Platforms," *The Vitality Institute*, January 2016, accessed at http://thevitalityinstitute.org/site/wp-content/uploads/2016/01/Vitality-HealthMetricsReportingRoadmap22Jan2016.pdf, accessed April, 2016.

In environmental health, early efforts focused on the impacts of corporate activities on the natural environment. However, corporate advocates of environmental sustainability were increasingly redefining their interest as creating "sustainable business," which meant improving the impact of corporate activities on people as well as on the environment.[21] Despite this progress, it was clear that business needed to create a new reporting category—or new scorecard altogether—to fully understand its population health footprint. In particular, businesses needed to develop better metrics for consumer health and community health, such that metrics across all four areas were denominated in the same "currency" and could be totaled.

[21] Harvard T. H. Chan School of Public Health, "Corporate Sustainability and Health (SHINE)," *SHINE website*, accessed at http://www.chgeharvard.org/category/corporate-sustainability-and-health-shine-0, accessed April 2016.

Creating an integrated measurement tool that spanned all four areas would enable a company to benchmark its current health activities and identify areas for improvement. Such an audit could reveal a surprising number of existing health-related activities. Simply communicating the totality of these disparate initiatives could provide a strong launching pad for a systematic Culture of Health program. As a first step to developing a benchmarking tool that could be used by any company, around 80 corporate executives attending an April 2016 Culture of Health conference were asked to complete a simple poll beforehand. The results, presented in Exhibit 6.6, indicated:

- Employee health was already the most important of the four pillars to corporate leaders and was also the most likely to be seen as related to positive financial returns.
- Within each of the four pillars, around 40 % of respondents agreed that there was much more their companies could do without sacrificing shareholder value.
- Fewer respondents agreed that their companies had innovative programs in place to support community and consumer health than employee and environmental health.

Exhibit 6.6: Culture of Health Conference Survey Questions and Results

In April 2016, the Culture of Health conference was held at Harvard Business School. A pre-conference survey was conducted to gauge the perspectives of participants working for corporations across four areas: Employee Health, Community Health, Consumer Health, and Environmental Health.

Respondents were asked to select how strongly they agreed (on a six-point scale) with the following statements. Results below are from 81 respondents.

Consumer Health

Statement:	Percent of Respondents who Selected "Agree" or "Strongly Agree"
Consumer health is very important to the leader of my corporation	62.0 %
My corporation has innovative programs and policies in place to support consumer health	50.6 %
Ensuring the healthfulness of our products and/or services is an important challenge for my corporation	58.2 %
There is much more that my corporation could do—without sacrificing shareholder value—to promote consumer health	36.7 %
I believe that improving consumer health can create positive financial returns for my corporation	59.0 %

Employee Health

Statement:	Percentage of Respondents who Selected "Agree" or "Strongly Agree"
Employee health is very important to the leader of my corporation	82.3 %
My corporation has innovative programs and policies in place to support employee health	60.8 %
Employers have a responsibility to ensure that their employees are healthy	72.2 %
Physical wellness is a challenge for a majority of employees at my corporation	29.1 %
Mental and family health is a challenge for a majority of employees at my corporation	30.4 %
There is much more that my corporation could do—without sacrificing shareholder value—to promote employee health	43.6 %
I believe that improving employee health can create positive financial returns for my corporation	87.3 %

Community Health

Statement:	Percent of Respondents who Selected "Agree" or "Strongly Agree"
Community health is very important to the leader of my corporation	60.8 %
My corporation has innovative programs and policies in place to support community health	49.4 %
For-profit corporations should invest more resources (money and human capital) to improve community health	57.0 %
For-profit corporations should create partnerships with non-profit and/or public actors to impact community health	69.2 %
There is much more that my corporation could do—without sacrificing shareholder value—to promote community health	43.0 %
I believe that improving community health can create positive financial returns for my corporation	66.7 %

Environmental Health

Statement:	Percent of Respondents who Selected "Agree" or "Strongly Agree"
Environmental health is very important to the leader of my corporation	72.2 %
My corporation has innovative programs and policies in place to support environmental health	59.5 %
Corporations have a responsibility to mitigate the risks posed by climate change	71.3 %
My corporation is transparent about its policies and its impacts on environmental health	58.2 %
There is much more that my corporation could do—without sacrificing share-holder value—to promote environmental health	35.4 %
I believe that improving environmental health can create positive financial returns for my corporation	62.5 %

Source: Casewriter

After auditing its existing population health footprint, a corporation had to then assess which health area to focus its resources on. Depending on its industry and company history, a company might find itself further down the path of progress in one area than another. Further, a company had to consider the level of resources to allocate to the Culture of Health vision and how to allocate these funds across multiple projects under the four pillars—all in a way that added up to a combined net positive health footprint. If the highest impact projects required longer-term investments, these sometimes had to be weighed against lower impact quick wins or proof of concept projects.

For example, a large retailer, like Target, had hundreds of stores serving individual communities. Its community engagement strategies were well-developed. Further, it sourced thousands of different products—presenting the large challenge of ensuring that they were all as healthful and safe for consumers as possible. On the other hand, Levi Strauss, a clothing manufacturer, focused on ensuring that its products were made with minimal environmental impact. Contributions to building community health were understandably more elusive because Levi Strauss, like all manufacturers who sold their products primarily through independent retailers or online, was one step removed from end consumers and their communities.

Guidance for PHF Metrics

In selecting metrics for calculating their population health footprints, corporations needed to consider and integrate both traditional business accounting metrics and common public health metrics (e.g., disability-adjusted life year[22]) in their reporting standards. Professors Robert Kaplan and David Norton or Harvard Business School, who devised the "balanced scorecard" in the 1990s, stated:

> What you measure is what you get. ... Executives also understand that traditional financial accounting measures like return on investment and earnings per share can give misleading signals for continuous improvement and innovation—activities that today's competitive environment demands....[23]

To be as useful as possible, population health footprint scorecards had to have these characteristics:

- **Complete**: Metrics should cover the full range of public health impacts of a corporation's activities.
- **Predictive**: Selected metrics should reveal causal relationships that could direct policy changes or further performance improvements.[24]
- **Measurable**: Whether quantitative or qualitative, metrics should be easy and unambiguous to measure.
- **Industry-specific**: To facilitate comparative benchmarking, metrics should be standardized by industry.[25]
- **Validated**: Metrics should be developed and their performance assessed by independent third parties. This was true, for example, of Colgate-Palmolive's Product Sustainability Scorecard, which evaluated products against 25 criteria in seven categories: responsible sourcing and raw materials, ingredient profile, water, social impact, packaging, energy and greenhouse gas, and waste.[26]

---

[22] Disability-Adjusted Life Year (DALY): One DALY can be thought of as one lost year of "healthy" life. The sum of these DALYs across the population, or the burden of disease, can be thought of as a measurement of the gap between current health status and an ideal health situation where the entire population lives to an advanced age, free of disease and disability (World Health Organization).

[23] Robert S. Kaplan and David P. Norton, "The Balanced Scorecard—Measures that Drive Performance," *Harvard Business Review*, January–February 1992, accessed at https://hbr.org/1992/01/the-balanced-scorecard-measures-that-drive-performance-2, accessed April 2016.

[24] Michael J. Mauboussin, "The True Measures of Success," *Harvard Business Review*, October 2012, accessed at https://hbr.org/2012/10/the-true-measures-of-success, accessed April 2016.

[25] Sustainability Accounting Standards Board (SASB), "Our Process," *SASB website*, http://www.sasb.org/approach/our-process/, accessed April 2016.

[26] John A. Quelch and Margaret L. Rodriguez, "Colgate-Palmolive Company: Marketing Anti-Cavity Toothpaste," HBS No. 9-515-050 (Boston: Harvard Business School Publishing, 2015).

Partnering for Change

Individual and population health were influenced by multiple factors, many beyond the control of corporations. Effectively pursuing a Culture of Health required companies to engage in partnerships and create shared value with other organizations from the public sector, private sector, academia, and civil society. Partnering skills were a core competency of companies committed to a Culture of Health, especially when it came to improving community health.

For example, among non-profit organizations that focused on health in innovative ways, Kaiser Permanente, a large integrated health system, introduced the idea of "total health" to its internal organization, patients and their families, and the surrounding communities.[27] It pursued workforce wellness initiatives for its employees, improved community access to healthy foods, advocated for increased physical activity in thousands of schools, and reduced its carbon footprint by purchasing green energy.[28]

Businesses could also work with governments to aid the success of health-focused government programs. Retailers, for example, could supply information on healthy behaviors and any health-related government programs directly to consumers through their many locations. In this respect, building a Culture of Health was not just for large, multinational companies with plentiful resources. Smaller, multi-generation family businesses that were embedded in their communities could, absent the pressure of quarterly reports to stockholders, find pursuing a Culture of Health motivating and rewarding.

Companies could also learn from one another or pool their efforts. Often, early successes that others believed they could replicate or build upon could advance an initiative beyond one company. For example, twenty of the largest companies in the US, which had 4 million employees and family members combined, created the Health Transformation Alliance to improve health outcomes and reduce costs.[29] A 2016 press release stated:

> … today they have joined together to improve the way health care benefits will be purchased for employees in an effort to create better health care outcomes. Their goal is to break with existing marketplace practices that are costly, wasteful and inefficient, all of which have resulted in employees paying higher premiums, copayments and deductibles every year.[30]

---

[27] TE Kottke, M Stiefel, NP Pronk, "Well-Being in All Policies": Promoting Cross-Sectoral Collaboration to Improve People's Lives, *Prev Chronic Dis*, 2016;13:160155, accessed at http://www.cdc.gov/pcd/issues/2016/16_0155.htm, accessed April 2016.

[28] Ibid.

[29] Health Transformation Alliance, "About," *Health Transformation Alliance website*, http://www.htahealth.com/, accessed April 2016.

[30] Health Transformation Alliance, Leading US Companies Announce Plan to Transform the Corporate Health Care System, *Health Transformation Alliance press release*, February 5, 2016, http://www.htahealth.com/docs/press/Leading_US_Companies_Announce_Plan_to_Transform_the_Corporate_Health_Care_System.pdf, accessed April 2016.

Some of these collaborative efforts had already seen impactful outcomes. The Healthy Weight Commitment Foundation (HWCF), a non-profit organization focused on reducing obesity—especially childhood obesity—had a coalition of 300 corporate and not-for-profit partners. In 2007, many of its corporate members—including 16 of the nation's leading food and beverage manufacturers—voluntarily pledged to redesign their product portfolios in order to collectively sell 1 trillion fewer calories in the US marketplace by 2012 and 1.5 trillion fewer calories by 2015.[31] A 2014 paper on the outcomes of the initiative found that, "The 16 HWCF companies collectively sold approximately 6.4 trillion fewer calories (−10.6%) in 2012 than in the baseline year of 2007."[32]

Another notable example came from Dow Chemical Company ("Dow"), a multinational chemical supply company headquartered since 1897 in Midland, Michigan, a town of about 42,000 people.[33,34] With more than 6000 employees in central Michigan, Dow had a significant stake in the health of those communities.[35] In the late 2000s, Dow partnered with the Michigan Health Improvement Alliance ("MiHIA").[36] A multi-stakeholder group comprised of private, public, and non-profit partners, MiHIA was a non-profit organization that worked to improve health delivery and outcomes in central Michigan. Partnership was at the core of the program, and MiHIA stated, "the concept of creating a multi-stakeholder collaborative was a novel one for the region and required building goodwill, gaining trust, articulating the value proposition and establishing a unifying process for all stakeholders."[37]

By 2015, Dow had taken on a leadership role within the partnership, with Dr. Catherine Baase, Dow's Chief Health Officer, as the Chairperson of the MiHIA board.[38] In January 2015, she testified to the US Senate:

---

[31] Shu Wen Ng, Meghan M. Slining, and Barry M. Popkin, "The Healthy Weight Commitment Foundation Pledge Calories Sold from U.S. Consumer Packaged Goods, 2007–2012," *Am J Prev Me*, 2014 October; 47(4): 508–519, accessed at http://www.ncbi.nlm.nih.gov/pmc/articles/PMC4171694/, accessed April 2016.

[32] Ibid.

[33] Dow, "Dow in Michigan," *Dow website*, accessed at http://www.dow.com/michigan/, accessed November 2015.

[34] United States Census Bureau, "American FactFinder: Community Facts," *Census website*, accessed at http://factfinder.census.gov/faces/nav/jsf/pages/community_facts.xhtml, accessed November 2015.

[35] John Gallagher and Jewel Gopwani, "Many in Midland worried about Dow Chemical takeover bid—Deal could be biggest-ever buyout," *WZZM 13 ABC*, 2007, accessed at http://archive.wzzm13.com/news/article/71634/0/Many-in-Midland-worried-about-Dow-Chemical-takeover-bidDOUBLEHYPHEN-Deal-could-be-biggest-ever-buyout, accessed November 2015.

[36] Marjorie Paloma, "Want a Healthier Workforce? Investing in Community Health Can Pay Off," *Robert Wood Johnson Foundation*: *Culture of Health*, August 18, 2015, accessed at http://www.rwjf.org/en/culture-of-health/2015/08/want_a_healthierwor.html, accessed November 2015.

[37] MiHIA, "History and Background," *MiHIA website*, accessed at http://www.mihia.org/index.php/about-mihia/history-background, accessed November 2015.

[38] MiHIA, "Governance," *MiHIA website*, accessed at http://www.mihia.org/index.php/governance, accessed November 2015.

> We also recognize that in our pursuit of the goals of our health strategy, the communities within which we operate and the health situation of those communities can be a great asset and a multiplier to our efforts. We see the benefit of constructive collaboration with our communities.[39]

Dow also enjoyed employee health improvements as a result of its participation in MiHIA. After discovering an elevated rate of diabetes in central Michigan communities and among Dow employees, Dow worked with MiHIA to develop a diabetes program integrated into its employee health promotion efforts.[40] Subsequent research showed that the prevalence of chronic conditions among Dow employees had fallen to 17 % below that of peer companies, and Dow spent 17 % less on diabetes and other chronic conditions than the peer group.[41]

Best Practices

Across corporations that had already begun to adopt a Culture of Health, the following best practices could be identified:

Aligning Mission and Vision

- Create or alter corporate mission, vision, and values statement to prioritize health.
- Think of health as an investment, not an expense.
- Communicate health as a corporate priority to employees, suppliers, consumers, and communities.

Organizing and Incentivizing Change

- Ensure CEO leadership and commitment to a Culture of Health that can motivate others within the company to model his/her behavior.

---

[39] U.S. Senate Committee on Health, Education, Labor and Pensions, Employer Wellness Programs: Better Health Outcomes and Lower Costs, "TESTIMONY OF CATHERINE BAASE, M.D. ON BEHALF OF THE DOW CHEMICAL COMPANY AND AMERICAN BENEFITS COUNCIL," January 29, 2015.

[40] Marjorie Paloma, "Want a Healthier Workforce? Investing in Community Health Can Pay Off," *Robert Wood Johnson Foundation*: *Culture of Health*, August 18, 2015, accessed at http://www.rwjf.org/en/culture-of-health/2015/08/want_a_healthierwor.html, accessed November 2015.

[41] Ibid.

- Develop a network of line managers to champion the Culture of Health throughout the organization.
- Appoint a chief health officer to oversee and improve the company's population health footprint.
- Tie executive bonuses, in part, to improvements in the relevant areas of the company's population health footprint.
- Give employees permission and monetary incentives to develop new Culture of Health programs—the best of which can then be scaled throughout the organization.

Analyzing Health Impacts

- Audit the company's current performance on each of the four areas within the population health footprint.
- Benchmark population health footprint results against those of industry peers and over time.
- Debate internally which initiatives will best leverage the company's capabilities and resources and where the biggest net positive health impacts can occur.
- Broaden the annual sustainability report into a health impact report that covers all four areas of the company's population health footprint.

Creating New Programs

- Design several Culture of Health experiments to prove the concept. Since change moves at the speed of trust, focus first achieving change in on one or two work groups or divisions or geographies, led by credible executives passionate about building a Culture of Health.
- Create new programs that not only positively impact health, but also generate business returns—thus creating shared value.
- Pay attention to cultural differences, especially those between developed and developing countries, in extending programs throughout the company's international network.

Scaling the Opportunity

- Adopt an open source approach to a Culture of Health; share best practices and assessment tools with other companies—especially small businesses—and non-corporate partners.
- Press suppliers to also adopt the Culture of Health mindset, thus improving the company's return on its health investments.

Conclusion

In promoting the Culture of Health agenda, the Robert Wood Johnson Foundation sought to improve the health of all Americans—in partnership with any and all who could help achieve this objective. Since only 20 % of health outcomes in the United States could be attributed to the health care sector, there was plenty of room for corporations, large and small, whatever industry they were in, to make a contribution.

All corporations laid down a population health footprint, impacting public health through their policies in the areas of consumer health, employee health, community health, and environmental health. By 2016, some corporations, such as Royal Caribbean and Target, had acknowledged their impact on—and responsibility for improving—public health. These corporations understood the business opportunities presented by health-related programs (e.g., increased revenues, reduced costs, improved brand reputation) and, as a result, they had aimed to create Cultures of Health—cultures where health was considered not only in the realms of regulatory compliance and risk mitigation, but as a core part of their business strategies. Within these organizations, all employees, from executives and managers to front-line employees, had to consider the health implications of their daily decisions.

However, most corporations had not considered their impacts on public health, or were just in the early stages of building a Culture of Health. As of 2016, few companies integrated their health-related efforts across departments. Even at corporations where health was prioritized, measuring the net impact of a company's population health footprint was rare. Businesses needed to adjust organizational structures, adjust their missions, and set new incentives so that their net impacts on public health were both measured and as positive as possible.

Of course, changes beyond corporate cultures were also necessary. Companies—especially smaller ones—needed simple software tools to enable them to conduct self-assessments of their current performance across the four pillars of the population health footprint and track improvements over time. Governments also needed to set an example by embracing the Culture of Health philosophy, providing incentives to corporations that supported health and integrating their own health-related initiatives across the departments (e.g., Health and Human Services, Agriculture, Energy, Environmental Protection Agency, Food and Drug Administration, and Consumer Product Safety Commission). Consumers needed to support organizations that prioritized health.

Maintaining and improving individual and population health was a monumental challenge. Corporations had a significant role to play. The Culture of Health and the population health footprint offered exciting new tools to enable business leaders to do their part, for the benefit of their shareholders as well as society at large.

About the Authors

John A. Quelch is the Charles Edward Wilson Professor of Business Administration at Harvard Business School and Professor in Health Policy and Management at Harvard TH Chan School of Public Health.

He is known for his work on prevention including a seminal 1980 article in Milbank Memorial Fund Quarterly "Marketing Principles and the Future of Preventive Health Care." He first testified on this topic before Congress at the age of 27. His many (co)authored books include *Problems and Cases in Health Care Marketing* (McGraw Hill, 2004), *Business Solutions for the Global Poor* (Jossey Bass, 2006) and *Consumers, Corporations and Public Health* (Oxford University Press, 2016).

Professor Quelch currently serves as a director of Alere, a health diagnostics company, Aramark, a food service and facilities management company, and Luvo, an innovative nutrition-focused food company.

Professor Quelch was educated at Oxford University (B.A.), the Wharton School of the University of Pennsylvania (M.B.A.), the Harvard School of Public Health (S.M.) and Harvard Business School (D.B.A.).

Emily C. Boudreau is a Research Associate at Harvard Business School. She previously worked for The Advisory Board Company and has experience advising hospitals, health systems, and companies in the healthcare industry. In 2015, she led a pro bono consulting team for the National Center for Medical-Legal Partnership and helped design screening tools for use in medical settings.

Emily graduated magna cum laude with distinction from Cornell University where her research focused on health policy issues. Her thesis on infrastructure and change management within the Veterans Health Administration received the Sherman-Bennett Prize.

J.A. Quelch, E.C. Boudreau, *Building a Culture of Health*, SpringerBriefs in Public Health, DOI 10.1007/978-3-319-43723-1

Index

J.A. Quelch, E.C. Boudreau, *Building a Culture of Health*, SpringerBriefs in Public Health, DOI 10.1007/978-3-319-43723-1